Springer Series in Reliability Engineering

Series Editor

Hoang Pham, Department of Industrial and Systems Engineering, Rutgers University, Piscataway, NJ, USA

More information about this series at http://www.springer.com/series/6917

Anatoly Lisnianski · Ilia Frenkel ·
Lev Khvatskin

Modern Dynamic Reliability Analysis for Multi-state Systems

Stochastic Processes and the L_Z-Transform

Springer

Anatoly Lisnianski
The Reliability Department
Israel Electric Corporation Ltd, Haifa,
Israel and Center for Reliability and Risk
Management, SCE-Shamoon College
of Engineering
Beer Sheva, Israel

Ilia Frenkel
Center for Reliability and Risk Management
SCE-Shamoon College of Engineering
Beer Sheva, Israel

Lev Khvatskin
Center for Reliability and Risk Management
and Industrial Engineering and Management
Department
SCE-Shamoon College of Engineering
Beer Sheva, Israel

ISSN 1614-7839 ISSN 2196-999X (electronic)
Springer Series in Reliability Engineering
ISBN 978-3-030-52487-6 ISBN 978-3-030-52488-3 (eBook)
https://doi.org/10.1007/978-3-030-52488-3

This Springer imprint is published by the registered company Springer Nature Switzerland AG
The registered company address is: Gewerbestrasse 11, 6330 Cham, Switzerland

To my son Roma and his wife Rima
my daughter Masha and her husband Igal
my beautiful granddaughters Yael, Liora,
Noa, Gaya and Maya
—Anatoly Lisnianski

To my grandchildren Idan, Sofia, Anastasia,
Thomas and Liam
—Ilia Frenkel

To my granddaughter Tomer
—Lev Khvatskin

Preface

This book is considered as a dynamic analysis of multi-state system (MSS) with focus on reliability, availability and performability analysis of aging MSS, components importance changing (variety) during the time, initial conditions impact on these issues, etc.

Using traditional methods for solving such problems for real-world multi-state systems leads to explosion of number of states that should be analysed. In order to avoid this problem, a special type of transform for Markov stochastic process that is called L_Z-transform was suggested. By using this transform, the Universal Generating Function technique, which is widely used for steady-state MSS reliability analysis, may be extended to dynamic reliability analysis and applied to random processes. Recently there are many successive applications of L_Z-transform method to dynamic analysis of different real-world multi-state systems. L_Z-transform was applied to determining age replacement policy in MSS, to provide reliability, availability and performability analysis for different specific MSSs such as power systems, refrigerating systems, air conditioning systems, different technical systems in aviation and maritime engineering.

The aim of this book is to provide a comprehensive, up-to-date presentation of an MSS dynamic analysis by using L_Z-transform and inverse L_Z-transform based on current achievements in this field. For these purposes the book is logically divided into two parts. In the first part (Chaps. 1–3) theoretical issues are presented with corresponding methodological examples. The second part (Chaps. 4–8) presents the method of implementation to solving practical problems.

The authors anticipate that the book will be attractive for researchers and practical engineers in addressing issues related to reliability, availability and performability analysis. In addition, it will be a helpful textbook for graduate courses in applied mathematics, industrial, electrical, mechanical and computer engineering.

It should be noticed that it is impossible to describe all the achievements in the field in a single book. Naturally some interesting results remained outside of the book's scope. In such cases the authors provide the readers with the corresponding references.

The book is organized as follows.

Chapter 1 presents the generic MSS model as a set of models of stochastic processes for every MSS's element and the system structure function. There were described indices for MSS reliability, availability and performability assessment.
In Chap. 2 in order to make the book self-contained, we have presented to the reader a basic knowledge about Markov processes. It was considered a methodology of Markov models building for reliability, availability and performability assessment.
Chapter 3 presents the approach for computation reliability, availability and performability of MSS in dynamic modes. For this purpose, there were introduced and mathematically defined L_Z-transform and inverse L_Z-transform for discrete-state continuous-time Markov process. The technique of their application for computation of MSS reliability, availability and performability indices was described in detail. Corresponding methodological examples were presented.
Chapter 4 presents the applications of L_Z-transform method to a short-term analysis of power systems. Here availability and performability assessment for coal-fired power units as well as for combine cycle units were considered. By using inverse L_Z-transform for coal-fired power plant a short-term risk function and reliability evaluation and estimating of mean time to failure were performed. It is shown how these parameters may be used in power system dispatch for making important operating decisions in real time.
In Chap. 5, the L_Z-transform method is applied for the analysis of a redundant multi-state air conditioning system for chemical laboratory that must function under different cases of hood functioning. Its availability is investigated when in the system there are some aging elements.
In Chap. 6 a reserved cold-water supply system for a factory, specialized in producing raw materials for the plastics industry is considered. Water cooling systems often have aging components and so, one should consider in general a non-homogeneous Markov model because some transition rates (intensities) are being time-dependent. Such a model is complex—even in simple cases it has hundreds of states. Therefore, it is rather difficult to build the model and to solve the corresponding system of differential equations by using straightforward Markov method. So, in order to make a dynamic MSS reliability analysis L_Z-transform will be used.
In Chap. 7 the method for analysis of impact of changing different failures or repair rates in different elements in aging MSS is presented. There were introduced sensitivity measures that are useful for investigation of an aging MSS and presented the method for their evaluation, based on using L_Z-transform. The method was applied to sensitivity analysis of real water cooling system. As sensitivity measures there were proposed derivations of MSS life time with respect to restriction of availability, mean output performance and accumulated performance deficiency.

In Chap. 8 the method for computation of the Birnbaum B-availability importance measure for components in an aging MSS under minimal repair was considered. The method is based on the L_Z-transform. It was shown that dynamic B-availability importance assessment for aging MSS is especially important, because the relative importance of MSS's components is changing over the time and strongly depends on the system demand.

We would like to express our sincere gratitude and appreciation to our colleagues from The Israel Electric Corporation, Dr. G. Levitin, Dr. D. Elmakias, Dr. H. BenHaim and Dr. D. Laredo and our colleagues and students from Shamoon College of Engineering, Israel for providing a supportive and intellectually stimulating environment. We also are very thankful to Prof. Yi Ding from Zhejiang University, China for his collaboration and very interesting discussions.

We are especially thankful to our teachers and friends Prof. Ilya Gertsbakh and Prof. Igor Ushakov that, unfortunately, recently have left us. Their fruitful ideas have a great impact on our work. We hope that their ideas will be very helpful for reliability researchers and engineers for a long time in the future.

Finally, it is indeed our pleasure to work with editorial staff from Springer, who has assisted in the book publication.

Beer Sheva, Israel
May 2020

Anatoly Lisnianski
Ilia Frenkel
Lev Khvatskin

Contents

1 Generic Model of Multi-state System. Reliability, Availability and Performability in Dynamic Modes 1
1.1 Generic Multi-state System Model 2
1.2 Reliability, Availability and Performability of Multi-state System 7
1.2.1 Reliability Measures of Multi-state Systems 9
1.2.2 Availability Measures of Multi-state Systems 10
1.2.3 Performability Measures of Multi-state Systems 12
1.3 Summary 15
References 16

2 Stochastic Processes Methods for MSS Reliability, Availability and Performability Assessment 17
2.1 General Concepts of Stochastic Processes Theory 17
2.2 Markov Models: Discrete-Time Markov Chains 21
2.2.1 Basic Definitions and Properties 21
2.2.2 Computation of n-Step Transition Probabilities and State Probabilities 23
2.3 Markov Models: Continuous-Time Markov Chains 27
2.3.1 Basic Definitions and Properties 27
2.3.2 Markov Models for Evaluating Reliability of Multi-state Elements 33
References 50

3 L_Z-Transform and Inverse L_Z-Transform of a Discrete-State Continuous-Time Markov Process 53
3.1 L_Z-Transform 54
3.1.1 L_Z-Transform Definition 56
3.1.2 Existence and Uniqueness 57

3.1.3 L_Z-Transform's Properties . . . 59
3.1.4 L_Z-Transform Application to MSS Reliability Analysis . . . 64
3.2 Inverse L_Z-Transform . . . 69
3.2.1 Computational Procedure for Determining Inverse L_Z-Transform . . . 70
3.2.2 Determining Set of States and Set of Initial Conditions for Entire MSS . . . 71
3.2.3 Determining Matrix A . . . 71
3.2.4 The Method Application . . . 73
3.2.5 Calculation of Reliability Function and Mean Time to Failure . . . 76
3.2.6 Numerical Example . . . 77
3.3 Summary . . . 82
References . . . 83

4 Short-Term Availability, Performability and Reliability Analysis for Power Systems . . . 85
4.1 Short-Term Availability and Performability Evaluation for Coal Fired Power Station by Using L_Z-Transform . . . 85
4.1.1 Multi-state Model for a Single Coal Fired Power Generating Unit . . . 85
4.1.2 Short-Term Reliability Analysis for Power Station with Several Coal Fired Units . . . 87
4.1.3 Short-Term Analysis for Power Plant with Three Coal Fired Generating Units . . . 90
4.2 Short-Term Availability and Performability Analysis of Power Plants with Combined Cycle Units . . . 96
4.2.1 Multi-state Markov Model for a Single Combined Cycle Generating Unit and L_Z-Transform for Its Output Generating Capacity Process . . . 96
4.2.2 Multi-state Reliability Analysis for Power System, Consisting of Number Combine Cycle Generating Units . . . 100
4.2.3 Short-Term Availability and Performability Analysis for Power Station with Three Combine-Cycle Units. Case Study . . . 102
4.3 Short-Term Reliability and Risk Evaluation for Power System Using Inverse L_Z-Transform . . . 106
4.3.1 Inverse L_Z-Transform Application to Risk Function Evaluation for Power Station . . . 107
4.3.2 Risk Evaluation for Power Station with Several Coal-Fired Generating Units. Case Study . . . 111
4.4 Summary . . . 119
References . . . 120

5 The L_Z-Transform Application for Availability Assessment of Air Conditioning System for Chemical Laboratories 121
5.1 Description of Chemical Laboratory Air Conditioning System and Its Elements 122
5.1.1 Description of the System 122
5.1.2 Description of the System's Elements 124
5.2 Multi-state Models for an Air Conditioning System for a Chemical Laboratory 126
5.3 Calculation of the Availability Indices of an Air Conditioning System for a Chemical Laboratory 130
5.4 Summary 132
References 134

6 Availability and Performability of the Reserved Cold Water Supply System for a Raw Material Manufacturing for Plastics Industry 135
6.1 Multi-state Model of the Reserved Cold Water Supply System . . . 136
6.1.1 System's Description 136
6.1.2 Description of System's Elements 137
6.2 The L_Z-Transform Application for Calculation of the Availability and Performability Indices of the Cold Water Supply System 141
6.3 Calculation of the Reliability and Performability Indices of the Cold Water Supply System 145
6.4 Summary 147
References 147

7 Sensitivity Evaluation for an Aging Multi-state System 149
7.1 Sensitivity Analysis for Aging MSS by Using L_Z-Transform Method 150
7.2 Multi-state Models of the Water Cooling System for Magnetic Resonance Inspection (MRI) Plant 152
7.2.1 System's Description 152
7.2.2 Description of Sub-systems and System's Elements 153
7.2.3 The L_Z-Transform Application for Calculation of the Reliability Indices of the Water Cooling System . . . 156
7.3 Sensitivity Analysis of the Water Cooling System 159
7.4 Summary 162
References 162

8 Birnbaum Importance Assessment for Aging Multi-state Water Cooling System 163
8.1 Dynamic B-Availability Importance Assessment by Using the L_Z-Transform Method 163
8.2 B-Availability Importance Assessment for the Water Cooling System of Magnetic Resonance Inspection Equipment 164

8.2.1 System Description and Corresponding Reliability Data 164
8.2.2 Computation L_Z-Transforms for System's Elements and Sub-systems 167
8.2.3 L_Z-Transform for the Entire MRI Water Cooling System Availability 174
8.2.4 B-Availability Importance Calculation 175
8.3 Summary 179
References 179

Appendix: MATLAB Codes for Case Studies Calculations 181

Chapter 1
Generic Model of Multi-state System. Reliability, Availability and Performability in Dynamic Modes

All systems are designed to perform their intended tasks in a given environment. Many systems can perform their tasks with various distinctive levels of efficiency usually referred to as performance rates. Traditional binary-state system has only two states with different performance rates—perfect functioning state (UP) and complete failure state (DOWN). A system that can have a finite number of performance rates is called a multi-state state system (MSS).

The basic concepts of MSS reliability have been intensively developed from the mid of the 1970's when first works in this field appeared. Since that time a numerous research works have been published and additional experience has also been gathered form industrial settings. The first book devoted to MSS reliability was published in 2003 [9]. In this book one can find the state of the art in the field at this stage. For readers who need deeper knowledge about MSS reliability may be recommended more recent books [8, 12]. In the book [14] one can find multi-state failure analysis of spacecraft subsystems.

In this book we shall concentrate on dynamic issues in MSS reliability, availability and performability.

Some dynamic reliability methods have been developed during the last years, such as Monte Carlo (MC) simulation, Dynamic Event Tree (DET) and Monte Carlo Dynamic Event Tree (MCDET) methods. In the paper [13] one can find overview of these methods and its applications to a realistic emergency power system that may be treated as MSS. A dynamic approach to the stochastic modelling of reliability is particularly appropriate for systems whose component failure characteristics are affected by their degree of use, amount of load, or extent of stresses experienced. Such systems arise in load-sharing models, software reliability models, and in multivariate failure-time models in biostatistics [6].

As it was noted in [7, 17] dynamic behavior is typical for modern engineering and computing systems. Such important parameters as system load, stress levels, redundancy levels can be changing with time. Additional important factors, which should be taken into account are possible aging of system components and impact

A. Lisnianski et al., *Modern Dynamic Reliability Analysis for Multi-state Systems*, Springer Series in Reliability Engineering,
https://doi.org/10.1007/978-3-030-52488-3_1

of initial conditions to short-term reliability, availability and performability. Often in real world problems some system's components are not in perfect states at initial time instant. This factor often has a great impact to the entire system failure. All factors that were mentioned above have an *impact not only to MSS reliability, but to MSS availability and performability* [10, 11]. The above-mentioned methods DET, MCDET and others are not oriented to MSS performability assessment. For example, aging, initial conditions changing, amount of load (MSS demand), etc. can significantly change MSS's components importance and sensitivity, as well as mean expected performance of entire system, expected performance deficiency etc. Therefore, such impact should be analyzed and taken into account in MSS design stage and during its operation.

Generally stochastic processes methods, especially Markov method and Monte-Carlo simulation are used in order to assess MSS performability in dynamic modes [15, 16]. Straightforward application of these methods requires a building of corresponding models, which in MSS's case has a big (and often even enormous) number of states. Building of such models and their analysis are very hard work.

In the book we will focus on specific mathematical method that avoids these difficulties.

1.1 Generic Multi-state System Model

Now main issues of MSS reliability theory are relatively well established. To the readers that would like to get a deeper knowledge in the field may be recommended the following books [8, 12].

In order to analyze MSS behavior one has to know the characteristics of its elements. Any system element j can have k_j different states corresponding to the performance rates, represented by the set $\mathbf{g}_j = \{g_{j1}, g_{j2}, \ldots, g_{jk_j}\}$, where g_{ji} is the performance rate of element j in the state i, $i \in \{1, 2, \ldots, k_j\}$.

Note that performance is usually characterized by some physical parameter. For example, in power system it may be a power generating capacity that is rating by MW, for data processing system it may be data processing speed that is rating by MB/sec and so on.

The performance rate $G_j(t)$ of element j at any instant $t \geq 0$ is a random process that takes its values from the set $\mathbf{g}_j$: $G_j(t) \in \mathbf{g}_j$. Therefore, for the time interval [0,T], where T is the MSS operation period, the performance rate of element j is defined as a discrete-state continuous-time stochastic process.

The probabilities associated with the different states of the system element j at any instant t can be represented by the set

$$\mathbf{p}_j(t) = \{p_{j1}(t), p_{j2}(t), \ldots, p_{jkj}(t)\}, \quad (1.1)$$

where

$$p_{ji}(t) = \Pr\{G_j(t) = g_{ji}\} \tag{1.2}$$

Note that since the element states compose the complete group of mutually exclusive events (meaning that the element j can always be in one and only in one of k_j states) $\sum_{i=1}^{k_j} p_{ji}(t) = 1$ for any t $(0 \le t \le T)$.

Observe that the behavior of binary elements (elements with only total failures) can also be represented by performance distribution. Indeed, consider a binary element b with a nominal performance (performance rate corresponding to a perfect functioning state) g^* and the probability that the element is in the perfect functioning state $p(t)$. Assuming that the performance rate of the element in a state of complete failure is 0, one obtains its performance distribution as follows:

$$\mathbf{g}_b = \{0, g^*\}, \ \mathbf{p}_b(t) = \{1 - p(t), p(t)\}$$

The steady-state (long-term or stationary, $t \to \infty$) performance distribution can be represented graphically in the form of cumulative curves. In this representation, each value of performance x corresponds to the probability that the element provides a performance rate that is no less than this level: $\Pr\{G_j \ge x\}$. For comparison, the graphs representing the performance distribution of the binary element b and the multi-state element j with five different states are presented in Fig. 1.1.

When the MSS consists of n elements, its performance rates are determined by the performance rates of these elements. At each moment, the system elements have certain performance rates corresponding to their states. The state of the entire system is determined by the states of its elements. Assume that the entire system has K different states and that g_i is the entire system performance rate in state i $(i \in \{1, \ldots, K\})$. The entire MSS performance rate at time instant t is a random variable that takes values from the set $\{g_1, \ldots, g_K\}$.

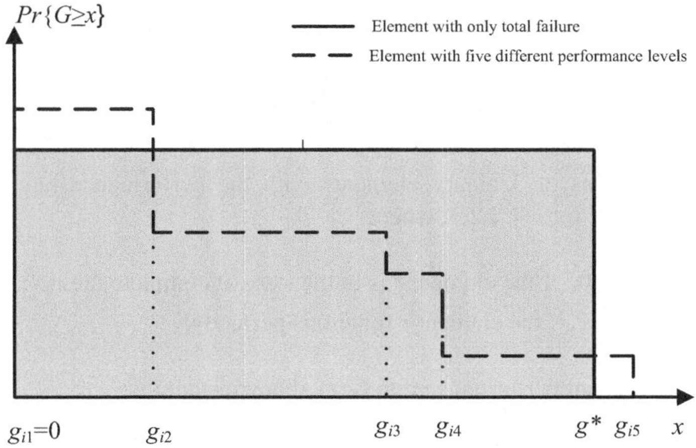

Fig. 1.1 Cumulative performance curves for steady-state behavior of multi-state elements

Definition Let $L^n = \{g_{11}, \ldots, g_{1k_1}\} \times \{g_{21}, \ldots, g_{2k_2}\} \times \cdots \times \{g_{n1}, \ldots, g_{nk_n}\}$ be a space of possible combinations of performance rates for all of the multi-state system elements and $M = \{g_1, \ldots, g_K\}$ is a space of possible values of the performance rate for the entire system.

The transform $f(G_1(t), \ldots, G_n(t)) : L^n \to M$, which maps the space of the elements' performance rates into the space of system's performance rates, is named the **MSS structure function**.

Note that the MSS structure function is an extension of a binary structure function. The only difference is in the definition of the state spaces: the binary structure function is mapped $\{0, 1\}^n \to \{0, 1\}$, while in the MSS, one deals with much more complex spaces.

Now we can define a generic model of the multi-state system.

The **generic MSS model** includes models of the performance stochastic processes,

$$G_j(t),\ j = 1, 2, \ldots, n \tag{1.3}$$

for each system element *j*, and the system structure function that produces the stochastic process corresponding to the output performance of the entire MSS,

$$G(t) = f(G_1(t), \ldots, G_n(t)). \tag{1.4}$$

In practice, performance stochastic processes $G_j(t)$ may be presented in some different forms. For example, performance probability distributions for all of the system elements may be given at any instant time t during the operation period $[0,T]$. Then the MSS is presented by these probability distributions $\boldsymbol{g}_j, \boldsymbol{p}_j(t), 1 \le j \le n$ and system structure function (1.4).

It also does not matter how the structure function is defined. It can be represented as a table, in analytical form, or be described as an algorithm for unambiguously determining the system performance $G(t)$ for any given set $f(G_1(t), \ldots, G_n(t))$. Below we'll consider examples for some possible representations of MSS structure functions.

Example 1.1

The system consists of 3 binary elements with the performance rates $G_i(t) \in \{g_{i1}, g_{i2}\} = \{0, 1\}$, for $i = 1,2,3$, where

$$g_{i1} = \begin{cases} 0, & \text{if the element } i \text{ is in the state of complete failure ;} \\ 1, & \text{if the element } i \text{ functions perfectly.} \end{cases}$$

The system output performance rate $G(t)$ at any instant t is:

Table 1.1 Structure function for 2-out of-3 system

$G_1(t)$	$G_2(t)$	$G_3(t)$	$f(G_1(t), G_2(t), G_3(t))$
0	0	0	0
0	0	1	0
0	1	0	0
0	1	1	1
1	0	0	0
1	0	1	1
1	1	0	1
1	1	1	2

$$G(t) = \begin{cases} 0, & \text{if there is more than one failed element ;} \\ 1, & \text{if there is only one failed element ;} \\ 2, & \text{if all the elements function perfectly.} \end{cases}$$

The values of the system structure function $G(t) = f(G_1(t), G_2(t), G_3(t))$ for all the possible system states are presented in Table 1.1.

Example 1.2
Consider a flow transmission system (Fig. 1.2a) consisting of three pipes [9]. The oil flow is transmitted from point C to point E. The pipes performance is measured by their transmission capacity (ton per minute). Elements 1 and 2 are binary. A state of total failure for both elements corresponds to a transmission capacity of 0 and the operational state corresponds to the capacities of the elements 1.5 and 2 ton per minute respectively so, that $G_1(t) \in \{0, 1.5\}$, $G_2(t) \in \{0, 2\}$. Element 3 can be in one of three states: a state of total failure corresponding to a capacity of 0, a state of partial failure corresponding to a capacity of 1.8 tons per minute and a fully operational state with a capacity of 4 tons per minute so that $G_3(t) \in \{0, 1.8, 4\}$. The system output performance rate is defined as the maximum flow that can be transmitted from C to E.

The total flow between points C and D through the parallel pipes 1 and 2 is equal to the sum of the flows through each of these pipes. The flow from point D to point E is limited by the transmitting capacity of element 3. On the other hand, this flow cannot be greater than the flow between points C and D. Therefore, the flow between points C and E (the system performance) is

$$G(t) = f(G_1(t), G_2(t), G_3(t)) = \min\{G_1(t) + G_2(t), G_3(t)\}.$$

The values of the system structure function $G(t) = f(G_1(t), G_2(t), G_3(t))$ for all the possible system states are presented in Table 1.2.

Example 1.3
Consider a data transmission system (Fig. 1.2b) consisting of three fully reliable network servers and three data transmission channels (elements). The data can be

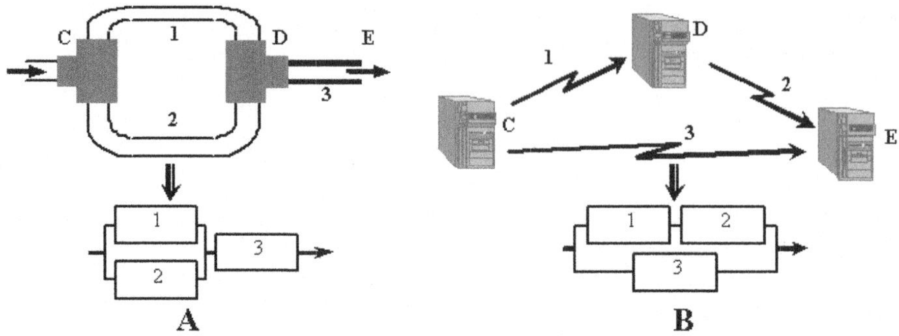

Fig. 1.2 Two different multi-state systems with identical structure functions

Table 1.2 Possible states of oil transmission system

$G_1(t)$	$G_2(t)$	$G_3(t)$	$f(G_1(t), G_2(t), G_3(t))$
0	0	0	0
0	0	1.8	0
0	0	4	0
0	2	0	1
0	2	1.8	1.8
0	2	4	2
1.5	0	0	0
1.5	0	1.8	1.5
1.5	0	4	1.5
1.5	2	0	0
1.5	2	1.8	1.8
1.5	2	4	3.5

transmitted from server C to server E through server D or directly. The time of data transmission between the servers depends on the state of the corresponding channel and is considered to be the channel performance rate. This time is measured in seconds.

Elements 1 and 2 are binary. They may be in a state of total failure when data transmission is impossible. In this case data transmission time is formally defined as ∞. They may also be in a fully operational state when they provide data transmission during 1.5 s and 2 s respectively: $G_1(t) \in \{\infty, 1.5\}$, $G_2(t) \in \{\infty, 2\}$. Element 3 can be in one of three states: a state of total failure, a state of partial failure with data transmission during 4 s and a fully operational state with data transmission during 1.8 s: $G_3(t) \in \{\infty, 4, 1.8\}$. The system performance rate is defined as the total time the data can be transmitted from server A to server C.

When the data is transmitted through server D, the total time of transmission is equal to the sum of times $G_1(t)$ and $G_2(t)$ it takes to transmit it from server C to server D and from server D to server E respectively. If either element 1 or 2 is in a state of total failure, data transmission through server D is impossible. For this case we formally state that $(\infty+2) = \infty$ and $(\infty+1.5) = \infty$. When the data is transmitted from server C to server E directly, the transmission time is $G_3(t)$. The minimum time needed to transmit the data from C to E directly or through D determines the system transmission time. Therefore, the MSS structure function takes the form:

$$G(t) = f(G_1(t), G_2(t), G_3(t)) = \min\{G_1(t) + G_2(t), G_3(t)\}.$$

Note that the different technical systems in Examples 1.2 and 1.3, even when having different reliability block diagrams (Fig. 1.2a, b), correspond to the identical MSS structure functions.

1.2 Reliability, Availability and Performability of Multi-state System

To numerically characterize MSS dynamic behavior one has to determine the MSS reliability, availability and performability indices. Here it should be noted that when we are speaking about performability, we are treating it as a *property* that characterizes a system capability to satisfy a customer demand. Therefore, as it was stated above, performance itself is usually a physical parameter and performability is a corresponding property that is measured as a customer demand satisfaction from system operation. So, when we are dealing with performability, we should consider a customer demand and determine measures for customer satisfaction. This approach was primarily declared in [4, 5]. Here we present its further developments.

From the customer point of view the entire set of possible system states usually can be divided into two disjoint subsets corresponding to acceptable and unacceptable states. The system entrance into the subset of unacceptable states constitutes a failure. MSS reliability can be defined as the system's ability to remain in acceptable states during the operational period.

Frequently the system state acceptability depends on the relation between the MSS performance and the desired level of this performance (demand) that is determined by the customer. In general, the demand $W(t)$ is also a random process. Below we shall consider the case when the demand is a discrete-state continuous-time stochastic process that takes discrete values from the set $\mathbf{w} = \{w_1, \ldots, w_M\}$. Often the customer desired relation between the system performance and demand can be expressed by the acceptability function $F(G(t), W(t))$ [9]. The acceptable states correspond to $F(G(t), W(t)) \geq 0$ and the unacceptable states correspond to $F(G(t), W(t)) < 0$. The last inequality defines the MSS failure criterion. In many practical cases, the MSS performance should exceed the demand. In such cases the acceptability function takes the form $F(G(t), W(t)) = G(t) - W(t)$.

Note that in some cases it may be impossible to divide an MSS state space into acceptable and unacceptable states. Only some functional associated with two stochastic processes G(t) and W(t) may be of interest in order to define undesired event or an MSS "failure". For example, MSS failure may be defined as an event when some functional $J = \int_0^T \alpha(G(t), W(t))dt$ greater than specified value J_0 and $\alpha(.)$ is defined as some arbitrary function. For a power system, where $G(t)$ and $W(t)$ are respectively treated as generating capacity and load (demand, which is required by consumers), functional J is interpreted as an energy not supplied to consumers. In this case $\alpha(t) = W(t) - G(t)$, if $W(t) - G(t) \geq 0$, and $\alpha(t) = 0$, if $W(t) - G(t) < 0$. Customer undesired event (failure) occurs when expected energy not supplied to consumers will be greater than some specified value.

Majority of the indices are based on considering the system evolution in the time domain. In this case the relation between the system's output performance and the demand represented by the two corresponding stochastic processes must be studied. Figure 1.3 shows an example of the behavior of the MSS performance and a demand as the realizations of two different independent stochastic processes.

When the system is considered in the given time instant or in a steady-state (when its output performance distribution does not depend on time) its behavior is determined by its performance represented as a random variable.

Note that in a steady state the distribution of the variable demand can be represented (in analogy with the distribution of the MSS performance) by two vectors $(\mathbf{w}, \mathbf{q})$, where $\mathbf{w} = \{w_1, \ldots, w_M\}$ is the vector of possible demand levels $w_j, j = 1, \ldots, M$ and $\mathbf{q} = \{q_1, \ldots, q_M\}$ is the vector of steady state probabilities of corresponding demand levels $q_j = Pr\{W = w_j\}, j = 1, \ldots, M$.

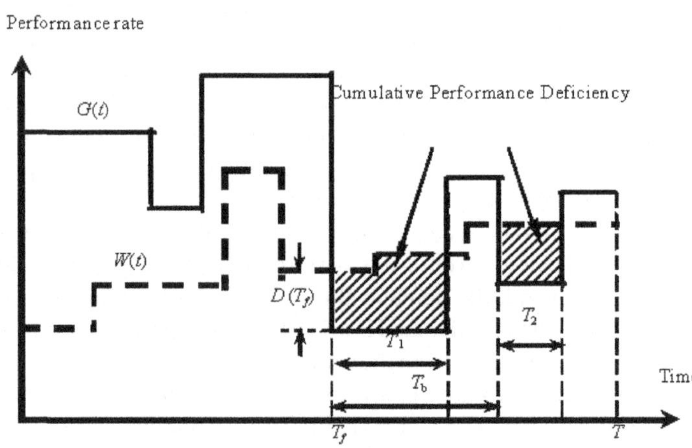

Fig. 1.3 MSS performance and demand as two stochastic processes

1.2.1 Reliability Measures of Multi-state Systems

When one considers an MSS evolution in the space of states during the system operation period T, the following random variables can be of interest:

- Time to failure, T_f is the time from the beginning of the system life up to the instant when the system enters the subset of unacceptable states the first time.
- Time between failures, T_b is the time between two consecutive transitions from the subset of acceptable states to the subset of unacceptable states.
- Number of failures, N_T is the number of times the system enters the subset of unacceptable states during the time interval $[0, T]$.

In Fig. 1.3, one can see an example of the random realization of two stochastic processes $G(t)$ and $W(t)$. Assume that the system performance value should exceed the value of demand: $F(G(t), W(t)) = G(t) - W(t)$. In this case, the first time that the process $G(t)$ down crosses the level of demand, $W(t)$ determines the time to MSS failure. This time is designated as T_f. The random variable T_f is characterized by the following indices:

- Probability of a failure-free operation or reliability function $R(t)$ is the probability that T_f is greater than or equal to the value t ($t > 0$), where in the initial state (at instant $t = 0$) MSS is in one of the acceptable states:

$$R(t) = \Pr\{T_f \geq t | F(G(0), W(0))\} \geq 0. \tag{1.5}$$

- Mean Time To Failure (MTTF) is the mean time up to the instant when the system enters the subset of unacceptable states for the first time:

$$E\{T_f\} \tag{1.6}$$

where $E\{\cdot\}$ is an expectation symbol.

The same two indices can be defined for the random variable T_b:

- The probability that the time between failures is greater than or equal to t:

$$\Pr\{T_b \geq t\} \tag{1.7}$$

- The Mean Time Between Failures (MTBF):

$$E\{T_b\} \tag{1.8}$$

The reliability indices associated with the random variable N_T are:

- The probability that N_T is not greater than some specified number n:

$$\Pr\{N_T \leq n\} \tag{1.9}$$

- The expected number of system failures in the interval $[0,T]$:

$$E\{N_T\} \tag{1.10}$$

Measures (1.9) and (1.10) are often important when logistic problems related to MSS operation are considered (for example, in order to determine the required number of spare parts).

1.2.2 Availability Measures of Multi-state Systems

MSS instantaneous (point) availability $A(t,w)$ is the probability that the MSS at instant $t > 0$ is in one of the acceptable states:

$$A(t, w) = \Pr\{F(G(t), W(t)) \geq 0\}. \tag{1.11}$$

MSS availability in the time interval $[0,T]$ is defined as:

$$A_T = \frac{1}{T}\int_0^T A(t, w)dt \tag{1.12}$$

The random variable A_T represents the portion of time when the MSS output performance rate is in an acceptable area. For example, in Fig. 1.10 $A_T = (T - T_1 - T_2)/T$. This index characterizes the portion of time when the MSS output performance rate is not less than the demand. The expected value of A_T is often used and is named as demand availability [2]:

$$A_D = E\{A_T\}. \tag{1.13}$$

Measures (1.11)–(1.13) characterize MSS availability.

In order to emphasize a difference between short-term and long-term (steady-state) measures we consider here long-term measures too.

For large t ($t \to \infty$), the system initial state has practically no influence on its availability. Therefore, the steady-state (stationary or long-term) MSS availability

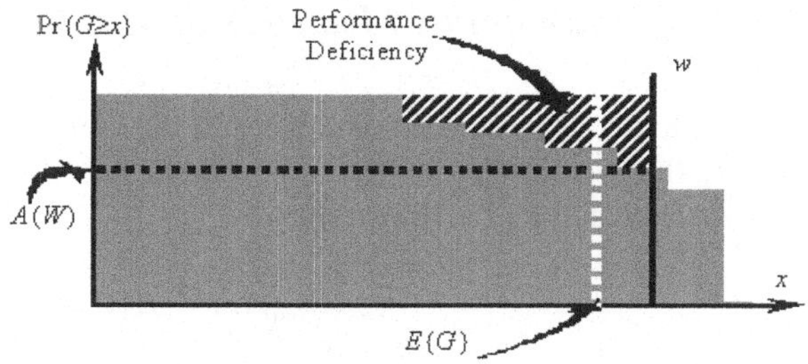

Fig. 1.4 MSS steady-state cumulative performance curve

$A_\infty(w)$ for the constant demand level $W(t) = w$ can be determined on the base of the system steady-state performance distribution:

$$A_\infty(w) = \sum_{k=1}^{K} p_k 1(F(g_k, w) \geq 0) \tag{1.14}$$

where $1(F(g_i, w) \geq 0) = \begin{cases} 1, & \text{if } F(g_i, w) \geq 0, \\ 0, & \text{if } F(g_i, w) < 0, \end{cases}$ and $p_k = \lim_{t\to\infty} p_k(t)$ is the steady-state probability of the MSS state k with the corresponding output performance rate g_k.

In the case where $F(G(t), W(t)) = G(t) - W(t)$, we have $F(g_k, w) = g_k - w$ and:

$$A_\infty(w) = \sum_{k=1}^{K} p_k 1(g_k \geq w) = \sum_{g_k \geq w} p_k. \tag{1.15}$$

Figure 1.4 demonstrates the cumulative performance curve for a MSS in a steady-state. In this figure the stationary availability corresponds to the point where the cumulative performance curve crosses the value of w.

When steady state demand is described by two vectors (***w***,***q***), the steady-state availability index may be obtained:

$$A_\infty(\mathbf{w}, \mathbf{q}) = \sum_{m=1}^{M} A(w_m) q_m = \sum_{m=1}^{M} q_m \sum_{k=1}^{K} p_k 1(F(g_k, w_m) \geq 0) \tag{1.16}$$

where $q_m = {T_m}\Big/{\sum_{m=1}^{M} T_m} = {T_m}/{T}$, $m = 1, 2, \ldots, M$.

In power engineering, the index $1 - A(t, w)$ is often used. This index is called the Loss of Load Probability (LOLP) [3]. The MSS performance in this case is interpreted as the power system generating capacity and its demand is interpreted as a load.

1.2.3 Performability Measures of Multi-state Systems

In order to obtain indices that characterize the average MSS output performance, one can use the performance expectation. The mean value of MSS instantaneous output performance at time t is determined as:

$$G_{mean}(t) = E\{G(t)\}. \tag{1.17}$$

If the long-run (steady-state) probabilities $p_k = \lim_{t\to\infty} p_k(t)$ exist, the steady-state expected performance takes the form:

$$G_\infty = \sum_{k=1}^{K} p_k g_k \tag{1.18}$$

The average MSS expected output performance for a fixed time interval $[0,T]$ is defined as:

$$G_T = \frac{1}{T}\int_0^T G_{mean}(t)dt \tag{1.19}$$

Observe that the mean MSS performance does not depend on demand.

In some cases a conditional expected performance is used. This index represents the mean performance of MSS under condition that it is in an acceptable state. In the steady-state it takes the form:

$$G_\infty^* = \sum_{k=1}^{K} g_k p_k 1(F(g_k, w) \geq 0) \Big/ A_\infty(w) \tag{1.20}$$

It is often important to know the measure of system performance deviation from a demand when the demand is not met. In the special case where $F(G(t), W(t)) = G(t) - W(t)$, the instantaneous performance deviation can be represented as:

$$D(t, w) = \max\{W(t) - G(t), 0\} \tag{1.21}$$

and is called the instantaneous performance deficiency at instant t. For example, in power systems $D(t)$ is interpreted as a generating capacity deficiency and determines the total power of consumers that must be immediately switched off from the system.

Since $D(t, w)$ is a random variable at time instant t, it can be characterized by the following measures:

- The probability that at instant t $D(t, w)$ does not exceed some specified level d:

$$\Pr\{D(t,w) \le d\} \tag{1.22}$$

- The mean value of the MSS performance deficiency (deviation) at instant t:

$$D_m(t,w) = E\{D(t,w)\} \tag{1.23}$$

When the MSS is in a steady-state and demand is constant $W(t) = w$, performance deficiency is not a function of time and can be obtained from the system steady-state performance distribution (Fig. 1.4 dashed area) as:

$$D_\infty = \sum_{k=1}^{K} p_k \max(w - g_k, 0) \tag{1.24}$$

For a variable demand represented by two vectors (**w**, **q**), steady state performance deficiency D_∞ takes the form:

$$D_\infty(\mathbf{w}, \mathbf{q}) = \sum_{m=1}^{M} \sum_{i=1}^{K} p_k q_m \max(w_m - g_i, 0) \tag{1.25}$$

The average MSS expected performance deficiency for a fixed time interval $[0, T]$ is defined as follows:

$$D_T = \frac{1}{T} \int_0^T D(t,w)dt \tag{1.26}$$

The cumulative performance deficiency for a fixed interval $[0,T]$ is defined as follows:

$$D_{\sum T} = \int_0^T D(t,w)dt. \tag{1.27}$$

For example, in power systems $D_{\sum T}$ is the energy not supplied to the consumers during the time interval $[0,T]$ (In Fig. 1.4 the cumulative performance deficiency is the sum of the dashed areas).

In some cases, the instantaneous performance deficiency makes no sense as the system uses storage facilities to accumulate a product. The deficiency appears not when the system performance does not meet the demand, but rather when the accumulated performance in interval $[0,T]$ is less than the accumulated demand at this interval. This takes place in oil and gas transmission systems with intermediate reservoirs. The accumulated performance deficiency in this case takes form:

$$D_{\sum T} = \int_{o}^{T} (W(t) - G(t))dt = \int_{0}^{T} W(t)dt - \int_{0}^{T} G(t)dt. \quad (1.28)$$

As $D_{\sum T}$ is a random variable, one can define the following characteristics:

- The probability that random variable $D_{\sum T}$ does not exceed some specified level ℓ (sometimes this measure is called throughput availability [2]:

$$\Pr\{D_{\sum T} \leq \ell\} \quad (1.29)$$

Note that in [1] was primarily presented mostly completed list of reliability measures for MSS.

- The expected amount of the product not supplied to consumers during the interval [0,T]:

$$D_{\sum m} = E\{D_{\sum T}\}. \quad (1.30)$$

Indices (1.18)–(1.30) characterize MSS performability.

Computation of steady-state indices is substantially less difficult than computation of short-term indices.

Computation of most of the above-mentioned indices is quite a difficult problem that is systematically studied in the following chapters of this book. For now, we are only considering a simple example of steady-state indices computation in order to illustrate some of measures introduced above.

Example 1.4

Consider two power system generators with a nominal capacity of 100 MW. In the first generator, some types of failures require the capacity to be reduced to 60 MW and other types lead to a complete outage. In the second generator, some types of failures require the capacity to be reduced to 80 MW, others lead to a capacity reduction to 40 MW and others lead to a complete outage.

The capacity and demand can be presented as a fraction of the nominal capacity. There are three possible relative capacity levels that characterize the performance of the first generator: $g_{11} = 0.0$, $g_{12} = 60/100 = 0.6$, $g_{13} = 100/100 = 1.0$ and four relative capacity levels that characterize the performance of the second generator: $g_{21} = 0.0$, $g_{21} = 40/100 = 0.4$, $g_{23} = 80/100 = 0.8$, $g_{23} = 100/100 = 1.0$.

Assume that the corresponding steady state probabilities are as follows: $p_{11} = 0.1$, $p_{12} = 0.6$, $p_{13} = 0.3$ for the first generator and $p_{21} = 0.05$, $p_{22} = 0.25$, $p_{23} = 0.3$, $p_{24} = 0.4$ for the second generator.

The required capacity level is 50 MW, which corresponds to $w = 50/100 = 0.5$.

The MSS stationary availability is:

$$A_{\infty 1}(w) = A_1(0.5) = \sum_{g_{1k} \geq 0.5} p_{1k} = 0.6 + 0.3 = 0.9,$$
$$A_{\infty 2}(w) = A_2(0.5) = \sum_{g_{2k} \geq 0.5} p_{2k} = 0.3 + 0.4 = 0.7.$$

The expected steady-state MSS performance (1.18) is:

$$G_{1\infty} = \sum_{k=1}^{3} p_{1k} g_{1k} = 0.1 \times 0 + 0.6 \times 0.6 + 0.3 \times 1.0 = 0.66$$

which means 66% of the nominal generating capacity for the first generator and

$$G_{2\infty} = \sum_{k=1}^{4} p_{2k} g_{2k} = 0.05 \times 0 + 0.25 \times 0.4 + 0.3 \times 0.8 + 0.4 \times 1.0 = 0.74$$

which means 74% of the nominal generating capacity for the second generator.

The steady-state performance deficiency (1.25) is:

$$D_{1\infty}(0.5) = \sum_{g_{1k} - W < 0} p_{1k}(w - g_{1k}) = 0.1 \times (0.5 - 0.0) = 0.05,$$
$$D_{2\infty}(0.5) = \sum_{g_{2k} - W < 0} p_{2k}(w - g_{2k}) = 0.05 \times (0.5 - 0.0) + 0.25 \times (0.5 - 0.4) = 0.05.$$

In this case, D_∞ may be interpreted as expected electric power unsupplied to consumers. The absolute value of this unsupplied demand is 5 MW for both generators. Multiplying this index by the considered system operating time T one can obtain the $D_{\sum T}$ index, which is interpreted as the expected unsupplied energy.

Note that since the obtained indices have different natures, they cannot be used interchangeably. For instance, in the present example, the first generator performs better than the second one when availability is considered ($A_1(0.5) > A_2(0.5)$), the second generator performs better than the first one when the expected capacity is considered ($G_{1\infty} < G_{2\infty}$), and both generators have the same unsupplied demand ($G_{1\infty} = G_{2\infty}$).

1.3 Summary

In this chapter was presented generic MSS model as a set of models of stochastic processes for every MSS's element and the system structure function. There were described indices for MSS reliability, availability and performability assessment. In the next chapter we consider evaluation of these indices in dynamic modes.

References

1. Aven T (1993) On performance measures for multistate monotone systems. Reliab Eng Syst Saf 41:259–266
2. Aven T, Jensen U (1999) Stochastic models in reliability. Springer, New York
3. Billinton R, Allan R (1996) Reliability evaluation of power systems. Plenum Press, New York
4. Boedigheimer R, Kapur K (1994) Customer-driven reliability models for multistate coherent systems. IEEE Trans Reliab 43:46–50
5. Brunelle R, Kapur K (1999) Review and classification of reliability measures for multi-state and continuum models. IIE Trans 31:1171–1180
6. Hollander M, Peña EA (1996) Dynamic reliability models. In: Jewell NP, Kimber AC, Lee ML, Whitmore GA (eds) Lifetime data: models in reliability and survival analysis. Springer, Boston, MA, pp 131–140
7. Labeau P, Smidts C, Swaminathan S (2000) Dynamic reliability: towards an integrated platform for probabilistic risk assessment. Reliab Eng Syst Saf 68(3):219–254
8. Lisnianski A, Frenkel I, Ding Y (2010) Multi-state system reliability analysis and optimization for engineers and industrial managers. Springer, London
9. Lisnianski A, Levitin G (2003) Multi-state system reliability. Assessment, optimization and applications. World Scientific, Singapore
10. Liu Y, Lin P, Huang HZ (2015) Bayesian reliability and performance assessment for multi-state systems. IEEE Trans Reliab 64(1):394–409
11. Misra K (2008) Performability engineering: an essential concept in the 21st century. In: Misra K (ed) Handbook of performability engineering. Springer, London, pp 1–12
12. Natvig B (2011) Multi-state systems reliability theory with applications. Wiley, New York
13. Pan X, Di Maio F, Zio E (2017) A benchmark of dynamic reliability methods for probabilistic safety assessment. In: Proceedings of the 2nd International Conference on System Reliability and Safety (ICSRS), Dec 20–22, 2017, Milan, Italy
14. Saleh J, Castet J-F (2011) Spacecraft reliability and multi-state failures. A statistical approach. Wiley, Chichester, UK
15. Trivedy K, Bobbio A (2017) Reliability and availability engineering. Modeling, analysis, and applications. Cambridge University Press, NY
16. Xie M, Dai Y-S, Poh K-L (2004) Computing system reliability. Models and analysis. Kluwer Academic, NY
17. Xing L, Levitin G, Wang Ch (2019) Dynamic system reliability. Modeling and analysis of dynamic and dependent behaviors. Wiley, NJ

Chapter 2
Stochastic Processes Methods for MSS Reliability, Availability and Performability Assessment

The purpose of this chapter is to describe basic concepts of applying a random process theory to MSS reliability assessment. Here, we do not present the basics of the measure-theoretic framework that are necessary to pure mathematicians. Readers who need this fundamental framework and more detailed stochastic processes presentation can find it in the following literature [3, 6, 8, 9, 13]. For reliability engineers and analysts, the books [4, 10, 11, 14] are especially recommended.

The Markov processes are widely used for reliability analysis because the number of failures in arbitrary time interval in many practical cases can be described as a Poisson process and the time up to the failure and repair time are often exponentially distributed. In the chapter detailed description of discrete-time Markov chain is presented as well as continuous-time Markov chain in order to provide for readers basic understanding of the theory and its engineering applications. It will be shown how by using the Markov processes theory MSS reliability measures can be determined. It will be also shown how such MSS reliability measures as the mean time up to the failure, mean number of failures in a time interval, mean sojourn time in a set of unacceptable states can be found by using the Markov reward models. These models also are the basis for unreliability associated costs assessment and life-cycle cost analysis.

2.1 General Concepts of Stochastic Processes Theory

A stochastic or random process is, essentially, a set of random variables, where the variables are ordered in a given sequence [12]. For example, the daily maximum temperatures at a weather station form a sequence of random variables, and this ordered sequence can be considered as a stochastic process. Another example is the sequence formed by the continuously changing number of people waiting in a queue at ticket window of a railway station.

A. Lisnianski et al., *Modern Dynamic Reliability Analysis for Multi-state Systems*, Springer Series in Reliability Engineering,
https://doi.org/10.1007/978-3-030-52488-3_2

More formally, the sequence of random variables in a process can be denoted by $X(t)$ where t is the index of the process.

The values assumed by the random variable $X(t)$ are called states, and the set of all possible values forms the state space of the process. So, a stochastic process is a sequence of random variables $\{X(t) | t \in T\}$, defined on a given probability space, indexed by the parameter t, where t varies over an index set T. In this book, we mainly deal with stochastic processes where t represents time.

A random variable X can be considered as the rule for assigning to every outcome ς of an experiment the value $X(\varsigma)$. A stochastic process is a rule for assigning to every ς the function $X(t, \varsigma)$. Thus, a stochastic process is a family of time functions depending on the parameter ς or, equivalently, a function of t and ς. The domain of ς is the set of all the possible experimental outcomes and the domain of t is a set of non-negative real numbers.

For example, an instantaneous speed of a car movement during its trip from point A to point B will be a stochastic process. A speed in each trip can be considered as an experimental outcome ς and each trip will have its own speed $X(t, \varsigma)$ that characterizes for this case an instantaneous speed of the trip as a function of time. This function will be different from such functions of other trips because of the influence of many random factors (such as wind, broad conditions etc.). In Fig. 2.1 one can see three different speed functions for three trips that can be treated as three different realizations of the stochastic process. It should be noticed that the cut of this stochastic process at time instant t_1 will present the random variable with mean V_m. In real-world systems many parameters such as temperature, voltage, frequency etc. may be considered as stochastic processes.

The time may be discrete or continuous. A discrete time may have a finite or infinite number of values; continuous time obviously has only an infinite number of values. The values taken by the random variables constitute the state-space. This state-space, in its turn, may be discrete or continuous. Therefore, stochastic processes may be classified into four categories according to whether their state-spaces and time are continuous or discrete. If the state space of a stochastic process is discrete, then it is called a discrete-state process, often referred to as a chain.

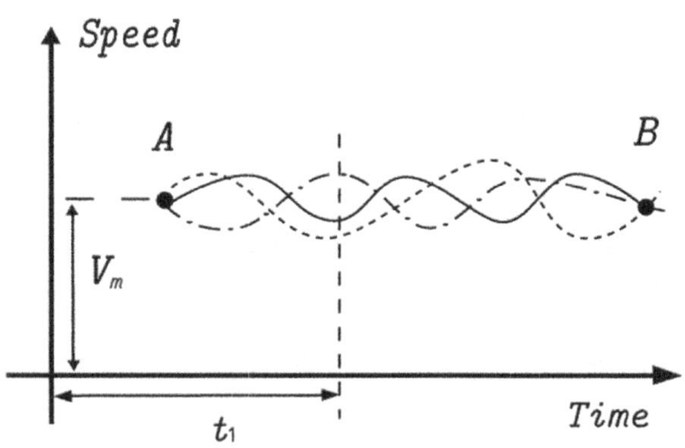

Fig. 2.1 Three realizations of stochastic process $V(t)$

The stochastic process $X(t, \varsigma)$ has the following interpretations:

1. It is a family of functions $X(t, \varsigma)$ where t and ς are variables.
2. It is a single time function or a realization (sample) of the given process if t is a variable and ς is fixed.
3. It is a random variable equal to the state of the given process at time t when t is fixed and ς is variable.
4. It is a number if t and ς are fixed.

One can use the notation $X(t)$ to represent a stochastic process omitting, as in the case of random variables, its dependence on ς.

For a fixed time $t = t_1$, the term $X(t_1)$ is a simple random variable that describes the state of the process at time t_1. For a fixed number x_1, the probability of the event $X(t_1) \leq x_1$ gives the CDF of the random variable $X(t_1)$, denoted by

$$F(x_1; t_1) = F_{X(t_1)}(x_1) = \Pr\{X(t_1) \leq x_1\} \tag{2.1}$$

CDF $F(x_1; t_1)$ is called the first-order distribution of the stochastic process $\{X(t)|t \geq 0\}$. Given two time instants t_1 and t_2, $X(t_1)$ and $X(t_2)$ are two random variables on the same probability space. Their joint distribution is known as the second order distribution of the process and is given by

$$F(x_1, x_2; t_1, t_2) = F_{X(t_1)X(t_2)}(x_1, x_2) = \Pr\{X(t_1) \leq x_1, X(t_2) \leq x_2\} \tag{2.2}$$

In general, the nth-order joint distribution of the stochastic process $\{X(t)|t \geq 0\}$ is defined by

$$\begin{aligned} F(x_1, x_2, \ldots, x_n; t_1, t_2, \ldots, t_n) &= F_{X(t_1)X(t_2)\ldots X(t_n)}(x_1, x_2, \ldots, x_n) = \\ &= \Pr\{X(t_1) \leq x_1, X(t_2) \leq x_2, \ldots, X(t_n) \leq x_n\} \end{aligned} \tag{2.3}$$

for all $t_1 < t_2 < \cdots < t_n$.

The last formula presents a complete description of a stochastic process. In practice to get such complete description of stochastic process is a very difficult task. Fortunately, in practice many stochastic processes permit a simpler description.

The simplest form of the joint distribution corresponds to a family of independent random variables. Then the joint distribution is given by the product of individual distributions.

Definition 2.1 A stochastic process $\{X(t)|t \geq 0\}$ is said to be an independent process if its nth-order joint distribution satisfies the condition:

$$F(x_1, x_2, \ldots, x_n; t_1, t_2, \ldots, t_n) = \prod_{i=1}^{n} F(x_i; t_i) = \prod_{i=1}^{n} \Pr\{X(t_i) \leq x_i\} \tag{2.4}$$

The assumption of an independent process considerably simplifies analysis, but it is often unwarranted and we are forced to consider some kind of dependence. The simplest and the very important type of dependence is the first-order dependence or Markov dependence.

Definition 2.2 A stochastic process $\{X(t)|t \geq 0\}$ is called a Markov process if for any $t_0 < t_1 < t_2 < \cdots < t_{n-1} < t_n < t$, the conditional distribution of $X(t)$ for given values of $X(t_0), X(t_1), \ldots, X(t_n)$ depends only on $X(t_n)$:

$$\begin{aligned} &\Pr\{X(t) \leq x|X(t_n) = x_n, X(t_{n-1}) = x_{n-1}, \ldots, X(t_1) = x_1, X(t_0) = x_0\} \\ &\quad = \Pr\{X(t) \leq x|X(t_n) = x_n\} \end{aligned} \tag{2.5}$$

This is general definition, which applies to Markov processes with continuous-state space. When MSS reliability is studied, discrete-state Markov processes or Markov chains are mostly concerned. In the next sections we will study both discrete-time and continuous-time Markov chains.

In the Markov process, the probabilities of the random variable at time $t > t_n$ depend on the value of the random variable at t_n but not on the realization of the process prior to t_n. In other words, the state probabilities at a future instant, given the present state of the process, do not depend on the states occupied in the past. Therefore, this process is also called "memoryless".

In many cases the conditional distribution (2.5) has the property of invariance with respect to the time origin t_n:

$$\Pr\{X(t) \leq x|X(t_n) = x_n\} = \Pr\{X(t - t_n) \leq x|X(0) = x_n\} \tag{2.6}$$

Such Markov process is said to be homogeneous.

In addition, we consider here two important stochastic processes that will be used in the future: point and renewal processes.

A point process is a set of random points t_i on the time axis. For each point process one can associate a stochastic process $X(t)$ equal to the number of points t_i in the interval $(0, t)$. In reliability theory point processes are widely used to describe the appearance of events in time (e.g., failures, terminations of repair, etc.)

An example of the point processes is the so-called *Poisson process.* The Poisson process is usually introduced using the Poisson points. These points are associated with certain events and the number $N(t_1, t_2)$ of the points in an interval (t_1, t_2) of length $t = t_2 - t_1$ is a Poisson random variable with parameter λt, where λ is the mean occurrence rate of the events:

$$\Pr\{N(t_1, t_2) = k\} = \frac{e^{-\lambda t}(\lambda t)^k}{k!} \tag{2.7}$$

If the intervals (t_1, t_2) and (t_3, t_4) are not overlapping, then the random variables $N(t_1, t_2)$ and $N(t_3, t_4)$ are independent. Using the points t_i, one can form the stochastic process $X(t) = N(0, t)$.

The Poisson process plays a special role in reliability analysis, comparable to the role of the normal distribution in probability theory. Many real physical situations can be successfully described with the help of Poisson processes.

A well-known type of point process is the so-called *renewal process*. This process can be described as a sequence of events, the intervals between which are independent and identically distributed random variables. In reliability theory, this kind of mathematical model is used to describe the flow of failures in time.

To every point process t_i one can associate a sequence of random variables y_n such that $y_1 = t_1$, $y_2 = t_2 - t_1, \ldots, y_n = t_n - t_{n-1}$, where t_1 is the first random point to the right of the origin. This sequence is called a renewal process. An example is the life history of the items that are replaced as soon as they fail. In this case, y_i is the total time the i-th item is in operation and t_i is the time of its failure.

One can see a correspondence among the following three processes:

- a point process t_i,
- a discrete-state stochastic process $X(t)$ increasing (or decreasing) by 1 at the points t_i,
- a renewal process consisting of the random variables y_i such that $t_n = y_1 + \cdots + y_n$.

A generalization of this type of process is the so-called alternating renewal process. This process consists of two types of independent and identically distributed random variables alternating with each other in turn. This type of process is convenient for the description of reparable systems. For such systems, periods of successful operation alternate with periods of idle time.

2.2 Markov Models: Discrete-Time Markov Chains

2.2.1 Basic Definitions and Properties

As it was described above, a Markov process is a stochastic process whose dynamic behavior is such that probability distribution for its future development depends only on the present state and not on how the process arrived in that state [7, 14]. When the state space, $\boldsymbol{S}$, is discrete (finite or countably infinite), then the Markov process is known as a Markov chain. Since the state space is discrete and countable we can assume without loss of generality that $S = \{0, 1, 2, 3, \ldots\}$. If the parameter space, $\boldsymbol{T}$, (remind that we usually will consider the time as the parameter) is discrete too, then we have a *discrete-time Markov chain*. Since the parameter space is discrete, we will let $T = \{0, 1, 2, 3, \ldots\}$. So, a Markov chain $\{X(n),\ n = 0, 1, 2, \ldots\}$ is described by a sequence of random variables $X(0) = x_0, X(1) = x_1, X(2) = x_2, \ldots$, where x_0, $x_1, x_2, \ldots$ are integer numbers. If the state of the system at time step n is j, we denote it as $X(n) = j$. Then X_0 is the initial state of the system at time step 0. By using these designations in analogy with (2.5), the Markov property can be defined as:

$$\Pr\{X_n = x_n | X_0 = x_0, X_1 = x_1, \ldots, X_{n-1} = x_{n-1}\} = \Pr\{X_n \le x_n | X_{n-1} = x_{n-1}\}. \tag{2.8}$$

As in the case of general Markov process, Eq. (2.8) implies that chain behavior in the future depends only on its present state and doesn't depend on its behavior in the past.

We designate the probability that at step n the chain will be in the state j as $p_j(n)$. So, we can write

$$p_j(n) = \Pr\{X_n = j\} \tag{2.9}$$

We also define the probability $p_{ij}(m, n)$ that the chain makes a transition to state j at step n, if at step m it was in state i. This probability is a conditional probability and we can write the following

$$p_{ij}(m, n) = \Pr\{X(n) = j | X(m) = i\},\ 0 \le m \le n. \tag{2.10}$$

Conditional probability $p_{ij}(m, n)$ is known as the *transition probability function* of the Markov chain.

Here we will only consider *homogeneous Markov chains*—those in which $p_{ij}(m, n)$ depends only on difference $n - m$. For such chains, the simpler notation

$$p_{ij}(n) = \Pr\{X(m+n) = j | X(m) = i\},\ 0 \le m \le n \tag{2.11}$$

is usually used to denote so-called *n-step transition probabilities.* In words, $p_{jk}(n)$ is the probability that a homogeneous Markov chain will move from state i to state j in exactly n steps.

If $n = 1$, for homogeneous Markov chain we also can write

$$p_{ij}(1) = \Pr\{X(m+1) = j | X(m) = i\} = p_{ij} = \text{const} \tag{2.12}$$

The probabilities p_{ij} are called one-step transition probabilities.

For engineering applications here we will consider only finite and countable state space $\mathbf{S} = \{0, 1, 2, \ldots, M\}$. The one-step transition probabilities can be condensed into a transition (one-step) probability matrix $\mathbf{P}$, where

$$\mathbf{P} = \left[p_{ij}\right] = \begin{bmatrix} p_{00} & p_{01} & \cdots & p_{0M} \\ p_{10} & p_{11} & \cdots & p_{1M} \\ \cdots & \cdots & \cdots & \cdots \\ p_{M0} & p_{M1} & \cdots & p_{MM} \end{bmatrix} \tag{2.13}$$

Since for all $i, j \in \mathbf{S}$, $0 \le p_{ij} \le 1$ and each row in $\mathbf{P}$ adds up to 1, the matrix $\mathbf{P}$ is a *stochastic matrix.*

The probability mass function of the random value $X(0)$ called the initial probability row-vector

$$\mathbf{p}(0) = \left[p_0(0), p_1(0), \ldots, p_M(0)\right] \tag{2.14}$$

and presents the initial conditions of Markov chain.

An equivalent description of the Markov chain can be given by a directed graph called the state-transition diagram (or state diagram for short) of the Markov chain. A node labeled i of the state diagram represents state i of the Markov chain and a branch labeled p_{ij} from node i to j represents corresponding one-step transition probability from state i to state j.

2.2.2 Computation of n*-Step Transition Probabilities and State Probabilities*

The problem considering here is in obtaining an expression for evaluating the n-step transition probability $p_{ij}(n)$ from the one-step transition probabilities $p_{ij} = p_{ij}(1)$. We remind that for homogeneous Markov chain according to expression (2.11) we have the following

$$p_{ij}(n) = \Pr\{X(m+n) = j | X(m) = i\},\ 0 \le m \le n.$$

Let's consider the transition probability $p_{ij}(m+n)$ that the process goes to state j at the $(m+n)$ step, given that at 0 step it is in state i. In order to reach state j at $(m+n)$ step the process firstly reaches some intermediate state k at step m with probability $p_{ik}(m)$ and then moves from k and reaches j at step $(m+n)$ with probability $p_{kj}(n)$. The Markov property implies that there are two independent events. Then using the theorem of total probability we obtain

$$p_{ij}(m+n) = \sum_k p_{ik}(m) p_{kj}(n) \tag{2.15}$$

Equation (2.15) is one form of the wide-known Chapman-Kolmogorov equation and provides efficient calculation of the n-step transitions probabilities.

We designate as $\mathbf{P}(n)$ the matrix of n-step probabilities or by another words, the matrix whose (i, j) entry is $p_{ij}(n)$. Then, if in (2.15) we let $m = 1$ and replace n by $n - 1$, we can rewrite Eq. (2.15) in matrix form:

$$\mathbf{P}(n) = \mathbf{P} \cdot \mathbf{P}(n-1) = \mathbf{P}^n \tag{2.16}$$

where $\mathbf{P}$ is one-step probabilities of the Markov chain.

In words, the n-step transition probabilities matrix is the nth power of the one-step transition probabilities matrix.

Based on the obtained results the unconditional state probabilities $p_j(n)$ can be examined. Their values depend on the initial state probabilities at $n = 0$ and on the number of steps passed since $n = 0$. It can be written:

$$p_j(n) = \Pr\{X(n) = j\} = \\ = \sum_i \Pr\{X(0) = i\} \Pr\{X(n) = j | X(0) = i\} = \sum_i p_i(0) p_{ij}(n). \tag{2.17}$$

In matrix form expression (2.17) can be rewritten as

$$\mathbf{p}(n) = \mathbf{p}(0) \cdot \mathbf{P}^n \tag{2.18}$$

where $\mathbf{p}(0)$ and $\mathbf{p}(n)$ are the row-vectors of the state probabilities initially (at step $n = 0$) and after the nth step respectively.

This implies that unconditional state probabilities of a homogeneous Markov chain are completely determined by the one-step transition probability matrix $\mathbf{P}$ and the initial probability vector $\mathbf{p}(0)$.

To illustrate the presented approach, we consider the following example.

Example 2.1 [1] Given a two-state Markov chain with the states denoted by 0 and 1 (Fig. 2.2).

The one-step transition probability matrix will be the following

$$\mathbf{P} = \begin{bmatrix} p_{00} & p_{01} \\ p_{10} & p_{11} \end{bmatrix} = \begin{bmatrix} 1 - p_{01} & p_{01} \\ p_{10} & 1 - p_{10} \end{bmatrix},$$

since it must hold that $p_{00} + p_{01} = 1$ and $p_{10} + p_{11} = 1$. Assume that $p_{01} = \alpha$ and $p_{10} = \beta$. Then

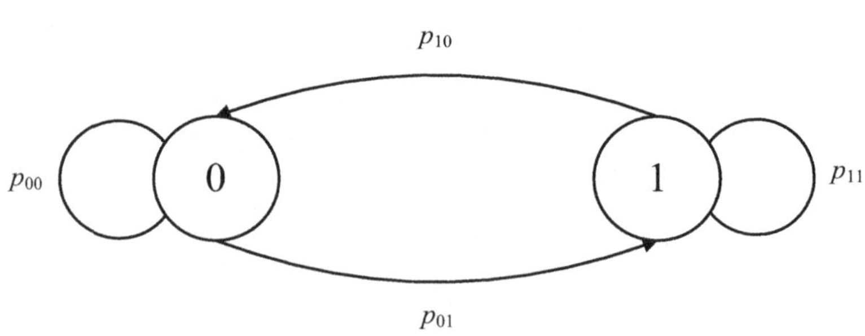

Fig. 2.2 Two-state discrete-time Markov chain

$$\mathbf{P} = \begin{bmatrix} 1-\alpha & \alpha \\ \beta & 1-\beta \end{bmatrix}.$$

The initial conditions at step $n = 0$ are the following $p_0(0) = a$ and $p_1(0) = 1-a$, so the initial state probabilities row-vector is given

$$\mathbf{p}(0) = \left[p_0(0), p_1(0)\right] = [a, 1-a].$$

Find n-step transition probability matrix and unconditional probabilities $p_0(n)$ and $p_1(n)$ of states 0 and 1 at step n respectively.

Solution According to the given one-step transition probability matrix **P** we can write

$$\begin{aligned} p_{00}(1) &= p_{00} = 1-\alpha, \; p_{01}(1) = p_{01} = \alpha, \\ p_{10}(1) &= p_{10} = \beta, \qquad p_{11}(1) = p_{11} = 1-\beta. \end{aligned}$$

For $n > 1$, using Eq. (2.16), we obtain

$$\begin{aligned} p_{00}(n) &= p_{00}(1)p_{00}(n-1) + p_{10}(1)p_{01}(n-1) = \\ &= (1-\alpha)p_{00}(n-1) + \beta p_{01}(n-1). \end{aligned}$$

Now since the row sums of matrix $\mathbf{P}^{n-1}$ are unity, we have

$$p_{01}(n-1) = 1 - p_{00}(n-1).$$

Substituting $p_{01}(n-1)$ to the previous equation we obtain for $n > 1$

$$p_{00}(n) = (1-\alpha)p_{00}(n-1) + \beta\left[1 - p_{00}(n-1)\right] = \beta + (1-\alpha-\beta)p_{00}(n-1).$$

By using the last recurrent equation, we can write the following

$$\begin{aligned} p_{00}(1) &= 1-\alpha, \\ p_{00}(2) &= \beta + (1-\alpha-\beta)(1-\alpha), \\ p_{00}(3) &= \beta + \beta(1-\alpha-\beta) + (1-\alpha-\beta)^2(1-\alpha), \\ &\ldots \\ p_{00}(n) &= \beta + \beta(1-\alpha-\beta) + \beta(1-\alpha-\beta)^2 + \cdots + \\ &\quad + \beta(1-\alpha-\beta)^{n-2} + (1-\alpha-\beta)^{n-1}(1-\alpha) = \\ &= \beta\left[\sum_{k=0}^{n-2}(1-\alpha-\beta)^k\right] + (1-\alpha-\beta)^{n-1}(1-\alpha). \end{aligned}$$

Based on the formula for the sum of a finite geometric series, we can write:

$$\sum_{k=0}^{n-2}(1-\alpha-\beta)^k=\frac{1-(1-\alpha-\beta)^{n-1}}{1-(1-\alpha-\beta)}=\frac{1-(1-\alpha-\beta)^{n-1}}{\alpha+\beta}.$$

Therefore, the expression for $p_{00}(n)$ can be rewritten in the following form

$$p_{00}(n)=\frac{\beta}{\alpha+\beta}+\frac{\alpha(1-\alpha-\beta)^n}{\alpha+\beta}.$$

Now $p_{01}(n)$ can be found

$$p_{01}(n)=1-p_{00}(n)=\frac{\alpha}{\alpha+\beta}-\frac{\alpha(1-\alpha-\beta)^n}{\alpha+\beta}.$$

Expressions for two remaining entries $p_{10}(n)$ and $p_{11}(n)$ can be found in a similar way. (Readers can do it themselves as an exercise).

Thus, the n-step transition probability matrix can be written

$$\mathbf{P}(n)=\mathbf{P}^n=\begin{bmatrix}\frac{\beta+\alpha(1-\alpha-\beta)^n}{\alpha+\beta} & \frac{\alpha-\alpha(1-\alpha-\beta)^n}{\alpha+\beta}\\ \frac{\beta-\beta(1-\alpha-\beta)^n}{\alpha+\beta} & \frac{\alpha+\beta(1-\alpha-\beta)^n}{\alpha+\beta}\end{bmatrix}.$$

Based on this n-step transition probability matrix and on the given initial state probabilities row-vector $\mathbf{p}(0)$, one can find state probabilities after n step by using formula (2.18)

$$\mathbf{p}(n)=\mathbf{p}(0)\mathbf{P}^n=[a,1-a]\cdot\begin{bmatrix}\frac{\beta+\alpha(1-\alpha-\beta)^n}{\alpha+\beta} & \frac{\alpha-\alpha(1-\alpha-\beta)^n}{\alpha+\beta}\\ \frac{\beta-\beta(1-\alpha-\beta)^n}{\alpha+\beta} & \frac{\alpha+\beta(1-\alpha-\beta)^n}{\alpha+\beta}\end{bmatrix}$$
$$=\left[\frac{\beta+(1-\alpha-\beta)^n}{\alpha+\beta}[a(\alpha+\beta)-\beta],\frac{\alpha-(1-\alpha-\beta)^n}{\alpha+\beta}[a(\alpha+\beta)-\beta]\right]$$

Therefore, state probabilities after n-step are the following

$$p_0(n)=\frac{\beta+(1-\alpha-\beta)^n}{\alpha+\beta}[a(\alpha+\beta)-\beta]$$
$$p_1(n)=\frac{\alpha-(1-\alpha-\beta)^n}{\alpha+\beta}[a(\alpha+\beta)-\beta]$$

2.3 Markov Models: Continuous-Time Markov Chains

2.3.1 Basic Definitions and Properties

As in the previous section, we confine our attention to discrete-state Markov stochastic processes or Markov chains. The continuous-time Markov chain is similar to that of the discrete-time case, except that the transitions from any given state to another state can take place at any instant of time. Therefore, for discrete-state continuous-time Markov chain the set of values $X(t)$ is discrete, $X(t) \in \{1, 2, \ldots\}$, and the parameter t has a continuous range of values, $t \in [0, \infty)$. In the reliability applications the set $\boldsymbol{S}$ of states is usually finite, $\boldsymbol{S} = \{1, 2, \ldots, K\}$ and so, $X(t) \in \{1, 2, \ldots, K\}$.

A discrete-state continuous-time stochastic process $\{X(t)|t \geq 0\}$ is called a Markov chain if for $t_0 < t_1 < \cdots < t_{n-1} < t_n$, its conditional probability mass function satisfies the relation

$$\begin{aligned} &\Pr\{X(t_n) = x_n | X(t_{n-1}) = x_{n-1}, \ldots, X(t_1) = x_1, X(t_0) = x_0\} \\ &= \Pr\{X(t_n) = x_n | X(t_{n-1}) = x_{n-1}\}. \end{aligned} \tag{2.19}$$

Introducing the notations $t = t_{n-1}$ and $t_n = t_{n-1} + \Delta t$ the expression (2.19) simplifies to:

$$\Pr\{X(t + \Delta t) | X(t) = j\} = \pi_{ji}(t, t + \Delta t) \tag{2.20}$$

The following designation is often used for the simplification

$$\pi_{ji}(t, t + \Delta t) = \pi_{ji}(t, \Delta t).$$

These conditional probabilities are called *transition probabilities*. If the probabilities $\pi_{ji}(t, \Delta t)$ do not depend on t, but only on the time difference Δt, the Markov process is said to be (*time-*) *homogeneous*. $\pi_{jj}(t, \Delta t)$ is the probability that no change in the state will occur in a time interval of length Δt given that the process is in state j at the beginning of the interval. Note, that

$$\pi_{ji}(t, t) = \begin{cases} 1, & \textit{if } j = i, \\ 0, & \textit{otherwise}. \end{cases} \tag{2.21}$$

Taking into account (2.21) one can define for each j a nonnegative continuous function $a_j(t)$:

$$a_j(t) = \lim_{\Delta t \to 0} \frac{\pi_{jj}(t, t) - \pi_{jj}(t, t + \Delta t)}{\Delta t} = \lim_{\Delta t \to 0} \frac{1 - \pi_{jj}(t, t + \Delta t)}{\Delta t} \tag{2.22}$$

and for each j and $i \neq j$ a nonnegative continuous function $a_{ji}(t)$:

$$a_{ji}(t) = \lim_{\Delta t \to 0} \frac{\pi_{ji}(t,t) - \pi_{ji}(t,t+\Delta t)}{\Delta t} = \lim_{\Delta t \to 0} \frac{\pi_{ji}(t,t+\Delta t)}{\Delta t} \tag{2.23}$$

The function $a_{ji}(t)$ is called the *transition intensity* from state i to state j at time t. For the homogeneous Markov processes, the transition intensities do not depend on t and therefore are constant.

If the process is in state j at a given moment, in the next Δt time interval there is either a transition from j to some state i or the process remains at j. Therefore

$$\pi_{jj}(\Delta t) + \sum_{i \neq j} \pi_{ji}(\Delta t) = 1. \tag{2.24}$$

Designating $a_{jj} = -a_j$ and combining (2.24) with (2.22) one obtains

$$a_{jj} = -a_j = \lim_{\Delta t \to 0} -\frac{1}{\Delta t} \sum_{i \neq j} \pi_{ji}(\Delta t) = -\sum_{i \neq j} a_{ji}. \tag{2.25}$$

Let $p_i(t)$ be the state probabilities of $X(t)$ at time t:

$$p_i(t) = \Pr\{X(t) = i\}, \; j = 1, \ldots, K; \; t \geq 0. \tag{2.26}$$

Expression (2.26) defines the probability mass function (pmf) of $X(t)$ at time t. Since at any given time the process must be in one of K states,

$$\sum_{i=1}^{K} p_i(t) = 1 \tag{2.27}$$

for any $t \geq 0$.

By using the theorem of total probability, for given $t > t_1$, we can express the pmf of $X(t)$ in terms of the transition probabilities $\pi_{ij}(t_1, t)$ and the pmf of $X(t_1)$:

$$\begin{aligned} p_j(t) = \Pr\{X(t) = j\} &= \sum_{i \in S} \Pr\{X(t) = j | X(t_1) = i\} \Pr\{X(t_1) = i\} \\ &= \sum_{i \in S} \pi_{ij}(t_1, t) p_i(t). \end{aligned} \tag{2.28}$$

If we let $t_1 = 0$ in (2.28), we obtain the following

$$p_j(t) = \sum_{i \in S} \pi_{ij}(0, t) p_i(0). \tag{2.29}$$

It means that the probabilistic behavior of a continuous-time Markov chain in the future is completely determined by the transition probabilities $\pi_{ji}(0, t)$ and the initial probability vector $\mathbf{p}(0) = [p_1(0), \ldots, p_K(0)]$.

The transition probabilities of a continuous-time Markov chain $\{X(t)|t \geq 0\}$ satisfy for all $i, j \in S$ the Chapman-Kolmogorov equation, which can be written for this case in the following form

$$\pi_{ij}(t_1, t) = \sum_{k \in S} \pi_{ik}(t_1, t_2)\pi_{kj}(t_2, t),\ 0 \leq t_1 < t_2 \leq t \tag{2.30}$$

The proof of this equation is based on the theorem of total probability:

$$\begin{aligned} &\Pr\{X(t) = j|X(t_1) = i\} \\ &= \sum_{k \in S} \Pr\{X(t) = j|X(t_2) = k, X(t_1) = i\} \Pr\{X(t_2) = k|X(t_1) = i\}. \end{aligned} \tag{2.31}$$

The subsequent application of the Markov property (2.20) to expression (2.31) yields (2.30).

The states probabilities at instant $t + \Delta t$ can be expressed based on states probabilities at instant t by using the following equations:

$$p_j(t + \Delta t) = p_j(t)\left[1 - \sum_{i \neq j} a_{ji}dt\right] + \sum_{i \neq j} p_i(t)a_{ij}dt,\ i, j = 1, \ldots, K. \tag{2.32}$$

Equation (2.32) can be obtained by using the following considerations.

The process can achieve the state j at instant $t + \Delta t$ by two ways.

1. The process may already be in the state j at instant t and doesn't leave this state up to the instant $t + \Delta t$. These events have probabilities $p_j(t)$ and $1 - \sum_{i \neq j} a_{ji}\Delta t$ respectively.
2. At instant t the process may be in one of the states $i \neq j$ and during time Δt transits from state i to state j. These events have probabilities $p_i(t)$ and $a_{ij}\Delta t$ respectively. These probabilities should be multiplied and summarized for all $i \neq j$ because the process can achieve state j from any state i.

Now one can rewrite (2.32) by using (2.29) and obtain the following

$$p_j(t + \Delta t) = p_j(t)\left[1 + a_{jj}\Delta t\right] + \sum_{i \neq j} p_i(t)a_{ij}\Delta t \tag{2.33}$$

or

$$\begin{aligned} p_j(t + \Delta t) - p_j(t) &= \sum_{\substack{i=1 \\ i \neq j}}^{K} p_i(t)a_{ij}\Delta t + p_j(t)a_{jj}\Delta t \\ &= \sum_{\substack{i=1 \\ i \neq j}}^{K} p_i(t)a_{ij}\Delta t - p_j(t)\sum_{\substack{i=1 \\ i \neq j}}^{K} a_{ji}\Delta t \end{aligned} \tag{2.34}$$

After dividing both sides of Eq. (2.34) by Δt and passing to limit $\Delta t \to 0$, we get

$$\frac{dp_j(t)}{dt} = \sum_{\substack{i=1 \\ i \neq j}}^{K} p_i(t)a_{ij} - p_j(t)\sum_{\substack{i=1 \\ i \neq j}}^{K} a_{ji},\ j = 1, 2, \ldots, K. \tag{2.35}$$

The system of differential equations (2.35) is used for finding the state probabilities $p_j(t)$, $j = 1, \ldots, K$ for the homogeneous Markov process when the initial conditions are given

$$p_j(t) = \alpha_j,\ j = 1, \ldots, K. \tag{2.36}$$

Mathematical details about (2.35) may be found in Trivedi [14] or in Ross [13].

When state-transition diagram for continuous-time Markov chain is built, Eq. (2.35) can be written by using the following rule: time-derivative of $p_j(t)$ for any arbitrary state j equals the sum of the probabilities of the states that have transitions to the state j multiplied by the corresponding transition intensities minus the probability of state j multiplied by the sum of the intensities of all transitions from the state j.

Introducing the row-vector $\mathbf{p}(t) = \left[p_1(t), p_2(t), \ldots, p_K(t)\right]$ and the transition intensity matrix $\mathbf{a}$

$$\mathbf{a} = \begin{bmatrix} a_{11} & a_{12} & \ldots & a_{1K} \\ a_{21} & a_{22} & \ldots & a_{2K} \\ \ldots & \ldots & \ldots & \ldots \\ a_{K1} & a_{K2} & \ldots & a_{KK} \end{bmatrix} \tag{2.37}$$

in which the diagonal elements are defined as $a_{jj} = -a_j$, we can rewrite the system (2.35) in matrix notation

$$\frac{dp(t)}{dt} = \mathbf{p}(t)\mathbf{a} \tag{2.38}$$

Note that the sum of the matrix elements in each row equals to 0: $\sum_{j=1}^{K} a_{ij} = 0$ for each i $(1 \leq i \leq K)$.

When the system state transitions are caused by failures and repairs of its elements, the corresponding transition intensities are expressed by the element's failure and repair rates.

The element's failure rate $\lambda(t)$ is the instantaneous conditional density of the probability of failure of an initially operational element at time t given that the element has not failed up to time t. Briefly, one can say that $\lambda(t)$ is the time-to-failure conditional probability density function (pdf). It expresses a hazard of failure in time instant t under a condition where there was no failure up to time t. The failure rate of an element at time t is defined as

$$\lambda(t) = \lim_{\Delta t \to 0} \frac{1}{\Delta t}\left[\frac{F(t+\Delta t) - F(t)}{R(t)}\right] = \frac{f(t)}{F(t)} \tag{2.39}$$

where $F(t)$ is the cdf of the time to failure of the element, $f(t)$ is pdf of the time to failure of the element, $R(t) = 1 - F(t)$ is the reliability function of the element.

For homogeneous Markov processes the failure rate doesn't depend on t and can be expressed as

$$\lambda = MTTF^{-1}, \tag{2.40}$$

where $MTTF$ is mean time to failure. Similarly, the repair rate $\mu(t)$ is the time-to-repair conditional pdf. For homogeneous Markov processes a repair rate does not depend on t and can be expressed as

$$\mu = MTTR^{-1}, \tag{2.41}$$

where $MTTR$ is the mean time to repair.

A state i is said to be an *absorbing state*, if once entered, the process is destined to remain in that state. A state j is said to be *reachable* from state i, if for some $t > 0$, $\pi_{ij}(t) > 0$. A continuous-time Markov chain is said to be *irreducible*, if every state is reachable from every other state.

In many applications, the long-run (final) or steady state probabilities $p_i = \lim_{t\to\infty} p_i(t)$ are of interest. For irreducible continuous-time Markov chain these limits are always exist for every state $i \in S$

$$p_i = \lim_{t\to\infty} p_i(t) = \lim_{t\to\infty} \pi_{ij}(t) = \lim_{t\to\infty} \pi_i(t) \tag{2.42}$$

and they are independent on the initial state $j \in S$. If the steady state probabilities exist, the process is called *ergodic*. For the final state probabilities, the computations become simpler. The set of differential equations (2.35) is reduced to a set of K algebraic linear equations because for the constant probabilities all time-derivatives are equal to zero, so, $\frac{dp_i(t)}{dt} = 0,\ i = 1, \ldots, K$.

Let the final states probabilities $p_i = \lim_{t\to\infty} p_i(t)$ exist. For this case in steady state, all derivatives of states probabilities in the right side of (2.35) will be zeroes. So, in order to find the long run probabilities the following system of algebraic linear equations should be solved

$$0 = \sum_{i=1}^{k} p_i(t) a_{ij},\ j = 1, 2, \ldots, K. \tag{2.43}$$

The K equations in (2.43) are not linearly independent (the determinant of the system is zero). An additional independent equation can be provided by the simple fact that the sum of the state probabilities is equal to 1 at any time:

$$\sum_{i=1}^{K} p_i = 1. \tag{2.44}$$

So, steady state probabilities of ergodic continuous-time Markov chain can be found by using expressions (2.43) and (2.44).

Now we consider additional important parameters of the process in steady state: state frequency and mean time of staying in state. The frequency f_i of state i, is defined as the expected number of arrivals into this state per unit time. Usually the concept of frequency is associated with the long-term (steady-state) behavior of the process. In order to relate the frequency, probability and mean time of staying in state i, we consider the system evolution in the state space as consisting of two alternating periods—the stays in i and the stays outside i. Thus, the process is represented by two states. Designate the mean duration of the stays in state i as $\bar{T}_i$ and that of the stays outside i, $\bar{T}_{oi}$. The mean cycle time, $\bar{T}_{ci}$ is then:

$$\bar{T}_{ci} = \bar{T}_i + \bar{T}_{oi} \tag{2.45}$$

From the definition of the state frequency it follows that, in the long run, f_i equals the reciprocal of the mean cycle time

$$f_i = \frac{1}{\overline{T}_{ci}} \tag{2.46}$$

Multiplying by $\bar{T}_i$ both two sides of Eq. (2.23) one gets

$$\overline{T}_i f_i = \frac{\overline{T}_i}{\overline{T}_{ci}} = p_i \tag{2.47}$$

Therefore,

$$f_i = \frac{p_i}{\overline{T}_i} \tag{2.48}$$

This is a fundamental equation, which provides the relation between the three state parameters in the steady state.

Unconditional random value T_i is minimal from all random values T_{ij} that characterize the conditional random time of staying in state i, if the transition will be performed from state i to any state $j \neq i$

$$T_i = \min\{T_{i1}, \ldots, T_{ij}\}. \tag{2.49}$$

All conditional times T_{ij} are distributed exponentially with the following cumulative distribution functions $F_{ij}(T_{ij} \leq t) = 1 - e^{-a_{ij}t}$. All transitions from state i are independent and, therefore, the cumulative distribution function of unconditional time T_i of staying in state i can be computed as the follows

$$F_i(T_i \leq t) = 1 - \Pr\{T_i > t\} = 1 - \prod_{j \neq i} \Pr\{T_{ij} > t\} =$$
$$= 1 - \prod_{j \neq i} \left[1 - F_{ij}(T_{ij} \leq t)\right] = 1 - \prod_{j \neq i} e^{-a_{ij}t} = 1 - e^{-\sum_{j \neq i} a_{ij}t}. \quad (2.50)$$

It means that unconditional time T_i is distributed exponentially with parameter $a_i = \sum_j a_{ij}$, and the mean time of staying in state i is the following

$$\overline{T}_i = \frac{1}{\sum_{j \neq i} a_{ij}} \quad (2.51)$$

Substituting $\bar{T}_i$ in the expression (2.48) we finally get

$$f_i = p_i \sum_{j \neq i} a_{ij} \quad (2.52)$$

Once state probabilities, p_i or $p_i(t)$ have been computed, reliability measures are usually obtained as corresponding functions of these probabilities.

2.3.2 *Markov Models for Evaluating Reliability of Multi-state Elements*

According to the generic MSS model any system element j can have k_j different states corresponding to the performance rates, represented by the set $\boldsymbol{g}_j = \{g_{j1}, \ldots, g_{jk_j}\}$. The current state of the element j and, therefore, the current value of the element performance rate $G_j(t)$ at any instant t, are random variables. $G_j(t)$ takes values from $\boldsymbol{g}_j : G_j(t) \in \boldsymbol{g}_j$. Therefore, for the time interval [0,T], where T is the MSS operation period, the performance rate of element j is defined as a stochastic process. Note that we consider only the Markov process where the state probabilities at a future instant do not depend on the states occupied in the past.

In this subsection when we deal with a single multi-state element, we can omit the index j for the designation of a set of the element's performance rates. So, this set is denoted as $\boldsymbol{g} = \{g_1, \ldots, g_k\}$. We also assume that this set is ordered so that $g_{i+1} \geq g_i$ for any i.

The elements can be divided into two groups. Those elements that are observed only until they fail belong to the first group. These elements either cannot be repaired, or the repair is uneconomical, or only the life history up to the first failure is of interest. Those elements that are repaired upon failure and their life histories consist of operating and repair periods belong to the second group. In the following subsections, both groups are discussed.

2.3.2.1 Non-repairable Multi-state Element

As mentioned above, the lifetime of a non-repairable element lasts until its first entrance into the subset of unacceptable states. In general, the acceptability of element's state depends on the relation between the element's performance and the desired level of this performance (demand). The demand $W(t)$ is also a random process that takes discrete values from the set $\boldsymbol{w} = \{w_1, \ldots, w_M\}$. The desired relation between the system performance and the demand can be expressed by the acceptability function $F(G(t), W(t))$.

First, consider a multi-state element with only minor failures defined as failures that cause element transition from state i to the adjacent state $i-1$. In other words, minor failure causes minimal degradation of element performance. The state-space diagram for such an element is presented in Fig. 2.3.

The element evolution in the state-space is the only performance degradation which is characterized by the stochastic process $\{G(t)|t \geq 0\}$. The transition intensity for any transition from state i to state $i-1$ is $\lambda_{i,i-1}$, $i = 2, \ldots, k$.

When sojourn time in any state i (or in other words, the time up to a minor failure in state i) is exponentially distributed with parameter $\lambda_{i,i-1}$, the process is the continuous-time Markov chain. Moreover, it is the widely known *pure death process* [14]. Let's define the auxiliary discrete-state continuous time stochastic process $\{X(t)|t \geq 0\}$, where $X(t) \in \{1, \ldots, k\}$. This process is strictly associated

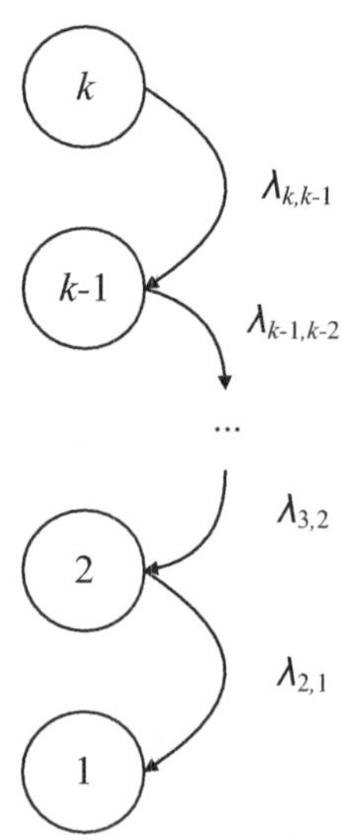

Fig. 2.3 State-transition diagram for non-repairable element with minor failures

with the stochastic process $\{G(t)|t \geq 0\}$. When $X(t) = i$, the corresponding performance rate of a multi-state element is g_i: $G(t) = g_i$. The process $X(t)$ is a discrete-state stochastic process decreasing by 1 at the points t_i, $i = 1, \ldots, k$, when the corresponding transitions occur. The state probabilities of $X(t)$ are

$$p_i(t) = \Pr\{X(t) = i\},\ i = 1, \ldots, k \tag{2.53}$$

for $t \geq 0$.

Note that

$$\sum_{i=1}^{k} p_i(t) = 1 \tag{2.54}$$

for any $t \geq 0$, since at any given time the process must be in some state.

According to the system (2.35), the following differential equations can be written in order to find state probabilities for the Markov process presented in Fig. 2.3:

$$\begin{cases} \dfrac{dp_k(t)}{dt} = -\lambda_{k,k-1} p_k(t) \\ \dfrac{dp_i(t)}{dt} = \lambda_{i+1,i} p_{i+1}(t) - \lambda_{i,i-1} p_i(t),\ i = 2, 3, \ldots, k-1 \\ \dfrac{dp_1(t)}{dt} = \lambda_{2,1} p_2(t) \end{cases} \tag{2.55}$$

One can see that in state k there is only one transition from this state to the state $k - 1$ with the intensity of $\lambda_{k,k-1}$ and there are no transitions to the state k. In each state i, $i = 2, 3, \ldots, k - 1$, there is one transition to this state from the previous state $i + 1$ with the intensity $\lambda_{i+1,i}$ and there is one transition from this state to state $i - 1$ with the intensity $\lambda_{i,i-1}$. Observe that there are no transitions from state 1. It means that if the process enters this state it is never left. The state 1 for non-repairable multi-state elements is the absorbing state.

We assume that the process begins from the best state k with a maximal element performance rate of g_k. Hence, the initial conditions are

$$p_k(0) = 1,\ p_{k-1}(0) = p_{k-2}(0) = \cdots = p_1(0) = 0 \tag{2.56}$$

Using widely available software tools, one can obtain the numerical solution of the system of differential equations (2.55) under initial conditions (2.56) even for large k. The system (2.55) can also be solved analytically by using Laplace-Stieltjes transform [5]. Using this transform and taking into account the initial conditions (2.56) one can represent (2.55) in the form of linear algebraic equations

$$\begin{cases} s\tilde{p}_k(s) - 1 = -\lambda_{k,k-1}\tilde{p}_k(s) \\ s\tilde{p}_i(s) = \lambda_{i+1,i}\tilde{p}_{i+1}(s) - \lambda_{i,i-1}\tilde{p}_i(s),\ i = 2, 3, \ldots, k-1 \\ s\tilde{p}_1(s) = \lambda_{2,1}\tilde{p}_2(s) \end{cases} \tag{2.57}$$

where $\tilde{p}_k(s) = L\{p_k(t)\} = \int_0^\infty e^{-st} p_k(t)$ is the Laplace-Stieltjes transform of a function $p_k(t)$ and $L\left\{\frac{dp_k(t)}{dt}\right\} = s\tilde{p}_k(s) - sp_k(0)$ is the Laplace-Stieltjes transform of the derivative of a function $p_k(t)$.

The system (2.57) may be rewritten in the following form

$$\begin{cases} \tilde{p}_k(s) = \dfrac{1}{s + \lambda_{k,k-1}}\tilde{p}_k(s) \\ \tilde{p}_i(s) = \dfrac{\lambda_{i+1,i}}{s + \lambda_{i,k-1}}\tilde{p}_k(s),\ i = 2, 3, \ldots, k-1 \\ \tilde{p}_1(s) = \dfrac{\lambda_{2,1}}{s}\tilde{p}_2(s) \end{cases} \tag{2.58}$$

Starting to solve this system from the first equation and sequentially substituting the obtained results to the next equation, one obtains

$$\begin{cases} \tilde{p}_k(s) = \dfrac{1}{s + \lambda_{k,k-1}} \\ \tilde{p}_i(s) = \dfrac{\lambda_{i+1,i}}{(s + \lambda_{i,i-1})}\dfrac{\lambda_{i+2,i+1}}{(s + \lambda_{i+1,i})}\cdots\dfrac{\lambda_{k,k-1}}{(s + \lambda_{k-1,k-2})}\dfrac{1}{(s + \lambda_{k,k-1})}, i = 2, 3, \ldots, k-1 \\ \tilde{p}_1(s) = \dfrac{\lambda_{2,1}}{s}\dfrac{\lambda_{3,2}}{(s + \lambda_{2,1})}\dfrac{\lambda_{4,3}}{(s + \lambda_{3,2})}\cdots\dfrac{\lambda_{k,k-1}}{(s + \lambda_{k-1,k-2})}\dfrac{1}{(s + \lambda_{k,k-1})} \end{cases} \tag{2.59}$$

Now in order to find the functions $p_k(t)$ the inverse Laplace-Stieltjes transform $L^{-1}\{\tilde{p}_k(s)\} = p_k(t)$ should be applied.

In the most common case when $F(g_i, w) = g_i - w$ (the element performance should not be less than the demand) for the constant demand level $g_{i+1} \geq w > g_i$ $(i = 1, \ldots, k-1)$ the acceptable states are the states $i + 1, \ldots, k$ where the element performance is above level g_i.

The probability of the state with the lowest performance $p_1(t)$ determines the unreliability function of the multi-state element for the constant demand level $g_2 \geq w > g_1$. Therefore, the reliability function defined as the probability that the element is not in its worst state (total failure) is

$$R_1(t) = 1 - p_1(t) \tag{2.60}$$

In general, if the constant demand is $g_{i+1} \geq w > g_i$, $(i = 1, \ldots, k-1)$, the unreliability function for the multi-state element is a sum of the probabilities of the unacceptable states 1, 2, …, i. So, the reliability function is

$$R_i(t) = 1 - \sum_{j=1}^{i} p_j(t) \quad (2.61)$$

The mean time up to multi-state element failure for this constant demand level can be interpreted as the mean time up to the process entering state i. It can be calculated as the sum of the time periods during which the process is remaining in each state $j > i$. Since the process begins from the best state k with the maximal element performance rate g_k (the initial conditions (2.56)), we have

$$MTTF_i = \sum_{j=i+1}^{k} \frac{1}{\lambda_{j,j-1}}, \; i = 1, 2, \ldots, k-1 \quad (2.62)$$

According to (1.17) one can obtain the element mean instantaneous performance at time t as

$$E_t = \sum_{i=1}^{k} g_i p_i(t) \quad (2.63)$$

The element mean instantaneous performance deficiency for the constant demand w according to (1.23) is

$$D_t = \sum_{i=1}^{k} p_i(t) \max(w - g_i, 0) \quad (2.64)$$

Example 2.2 As a non-repairable multi-state element that can have only minor failures we consider an electric generator installed in the airplane where its maintenance is impossible during the flight [2]. The generator has $k = 4$ possible performance levels (in the states 4, 3, 2 and 1 its capacities are $g_4 = 10$ KW, $g_3 = 8$ KW, $g_2 = 5$ KW and $g_1 = 0$ respectively) and the following failure rates: $\lambda_{4,3} = 2\,\text{year}^{-1}$, $\lambda_{3,2} = 1\,\text{year}^{-1}$ and $\lambda_{2,1} = 0.7\,\text{year}^{-1}$. The initial state is the best state 4.

Each flight duration $T_{flight} = 10$ h. The airplane was designed for $N_{flight} = 50$ flights up to general maintenance on the ground. Therefore, the service time up to the general maintenance is defined as $T_{service} = 500$ h.

We should find the expected energy not supplied to the airplane's consumers during the airplane service time, the probability that such failure as decreasing generation capacity lower than demand level 6 KW occurs during the service time and the mean time up to this failure

Solution In order to find state probabilities the following system of differential equations should be solved according to (*2.55*)

$$\begin{cases} \dfrac{dp_4(t)}{dt} = -\lambda_{4,3}p_4(t) \\ \dfrac{dp_3(t)}{dt} = \lambda_{4,3}p_4(t) - \lambda_{3,2}p_3(t) \\ \dfrac{dp_2(t)}{dt} = \lambda_{3,2}p_3(t) - \lambda_{2,1}p_2(t) \\ \dfrac{dp_1(t)}{dt} = \lambda_{2,1}p_2(t) \end{cases}$$

with the initial conditions $p_4(t) = 1, p_3(t) = p_2(t) = p_1(t) = 0$.

Using the Laplace-Stieltjes transform, we obtain

$$\tilde{p}_4(s) = \frac{1}{s + \lambda_{4,3}}, \ \tilde{p}_3(s) = \frac{\lambda_{4,3}}{(s + \lambda_{3,2})(s + \lambda_{4,3})},$$
$$\tilde{p}_2(s) = \frac{\lambda_{3,2}\lambda_{4,3}}{(s + \lambda_{2,1})(s + \lambda_{3,2})(s + \lambda_{4,3})}, \ \tilde{p}_1(s) = \frac{\lambda_{2,1}\lambda_{3,2}\lambda_{4,3}}{s(s + \lambda_{2,1})(s + \lambda_{3,2})(s + \lambda_{4,3})}.$$

Using the inverse Laplace-Stieltjes transform, we find the state probabilities as functions of time t:

$$\begin{aligned} p_4(t) &= e^{-\lambda_{4,3}t} \\ p_3(t) &= \frac{\lambda_{4,3}}{\lambda_{4,3}-\lambda_{3,2}}\left(e^{-\lambda_{3,2}t} - e^{-\lambda_{4,3}}\right) \\ p_2(t) &= \frac{\lambda_{3,2}\lambda_{4,3}[(\lambda_{4,3}-\lambda_{3,2})e^{-\lambda_{2,1}t}+(\lambda_{2,1}-\lambda_{4,3})e^{-\lambda_{3,2}t}+(\lambda_{3,2}-\lambda_{2,1})e^{-\lambda_{4,3}t}]}{(\lambda_{3,2}-\lambda_{2,1})(\lambda_{4,3}-\lambda_{3,2})(\lambda_{2,1}-\lambda_{4,3})} \\ p_4(t) &= 1 - p_2(t) - p_3(t) - p_4(t) \end{aligned}$$

These probabilities are presented in Fig. 2.4.

Now we can obtain the reliability measures for this multi-state element. The reliability functions for different demand levels are according to (2.61):

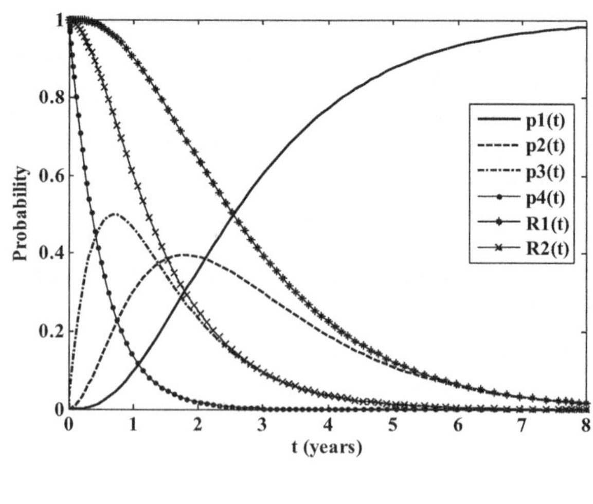

Fig. 2.4 State probabilities and reliability measures for unrepairable element with minor failures

$$R_1(t) = 1 - p_1(t), \text{ for } g_1 < w \le g_2,$$
$$R_2(t) = 1 - p_1(t) - p_2(t), \text{ for } g_2 < w \le g_3,$$
$$R_3(t) = 1 - p_1(t) - p_2(t) - p_3(t) = p_4(t), \text{ for } g_3 < w \le g_4.$$

These reliability functions are also presented in Fig. 2.4.

According to (2.64) we obtain the element mean instantaneous performance at time t

$$E_t = \sum_{k=1}^{4} g_k p_k(t) = 10p_4(t) + 8p_3(t) + 5p_2(t) + 0p_1(t)$$

The demand is constant during the flight and $w = 6$ KW. Therefore, according to (2.64), the element mean instantaneous performance deficiency is

$$D_t = \sum_{k=1}^{4} p_k(t)\max(w - g_k, 0) = 1p_2(t) + 6p_1(t)$$

Functions E_t, D_t are presented in the Fig. 2.5.

Note, that expected energy not supplied ($EENS$) to the airplane consumers during the service time $T_{service}$= 500 h will be the following

$$EENS = \int_0^{T_{service}} D_t dt \approx 0.547 \text{ KWh}.$$

Now based on (2.62) we obtain the mean time to failure

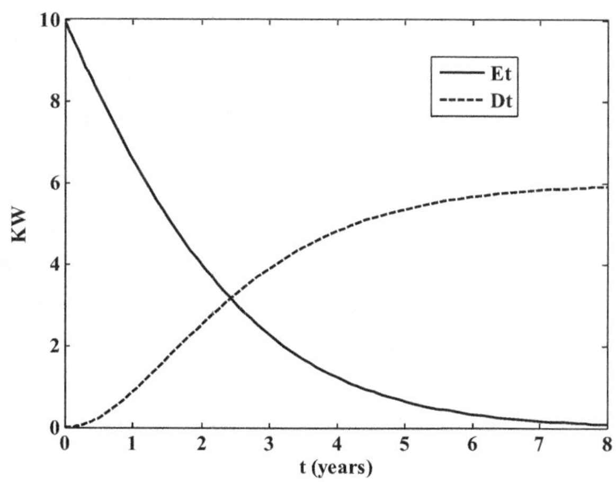

Fig. 2.5 Mean instantaneous performance and mean instantaneous performance deficiency for unrepairable element with minor failures

$$MTTF_1 = \frac{1}{\lambda_{4,3}} + \frac{1}{\lambda_{3,2}} + \frac{1}{\lambda_{2,1}} = 2.93 \text{ year for } g_1 < w \leq g_2,$$
$$MTTF_2 = \frac{1}{\lambda_{4,3}} + \frac{1}{\lambda_{3,2}} = 1.5 \text{ year for } g_2 < w \leq g_3,$$
$$MTTF_3 = \frac{1}{\lambda_{4,3}} = 0.5 \text{ year for } g_3 < w \leq g_4.$$

For the constant demand $w = 6$ KW, the mean time to failure is equal to $MTTF_2 = 1.5$ year. The probability that this failure (decreasing the generating capacity lower than demand level 6 KW) will not occurs during the service time according to the graph in Fig. 2.4 will be the following

$$R_2(t = T_{service}) = R_2(500h) = 0.997.$$

Now consider a non-repairable multi-state element that can have both minor and major failures (major failure is a failure that causes the element transition from state i to state j: $j < i - 1$). The state-space diagram for such an element representing transitions corresponding to both minor and major failures is presented in Fig. 2.6.

For the continuous-time Markov chain that is represented by this state-space diagram, the following system of differential equations for state probabilities can be written according to the Eqs. (2.35):

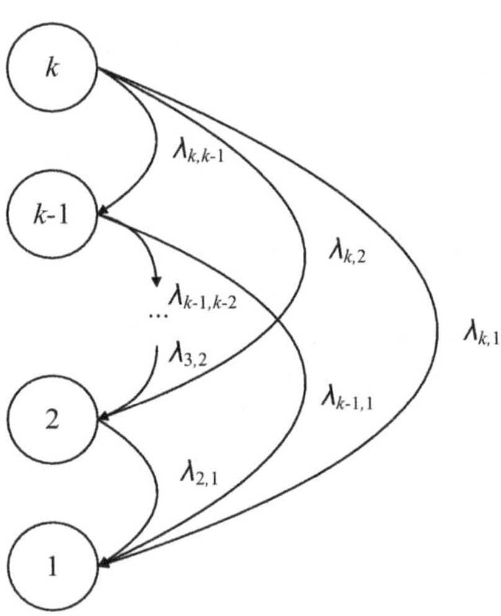

Fig. 2.6 State-transition diagram for non-repairable element with minor and major failures

$$\begin{cases} \dfrac{dp_k(t)}{dt} = -p_k(t)\displaystyle\sum_{e=1}^{k-1} \lambda_{k,e} \\ \dfrac{dp_i(t)}{dt} = \displaystyle\sum_{e=i+1}^{k} \lambda_{e,i}p_e(t) - p_i(t)\sum_{e=1}^{i-1} \lambda_{i,e}, i = 2, 3, \ldots, k-1 \\ \dfrac{dp_1(t)}{dt} = \displaystyle\sum_{e=2}^{k} \lambda_{e,1}p_e(t) \end{cases} \tag{2.65}$$

with the initial conditions (2.56).

After solving this system and obtaining the state probabilities $p_i(t)$, $i = 1, \ldots, k$ the mean instantaneous performance and the mean instantaneous performance deficiency can be determined by using (2.63) and (2.64).

As in the case of the unrepairable multi-state element with minor failures, the unavailability of the element with both minor and major failures is equal to the sum of the probabilities of unacceptable states. Therefore, for the constant demand w ($g_i < w \leq g_{i+1}$) one can use the expression (2.61) for determining the element reliability function.

The straightforward method for finding the mean time up to failure is not applicable for multi-state elements with minor and major failures.

2.3.2.2 Repairable Multi-state Elements

The more general model of a multi-state element is the model with repair. The repairs can also be both minor and major. The minor repair returns an element from state j to state $j + 1$ while the major repair returns it from state j to state i, where $i > j + 1$.

The special case of the repairable multi-state element is an element with only minor failures and minor repairs. The stochastic process corresponding to such an element is named the birth and death process. The state-space diagram of this process is presented in Fig. 2.7a.

The state-space diagram for the general case of the repairable multi-state element with minor and major failures and repairs is presented in Fig. 2.7b. The following system of differential equations can be written for the state probabilities of such element:

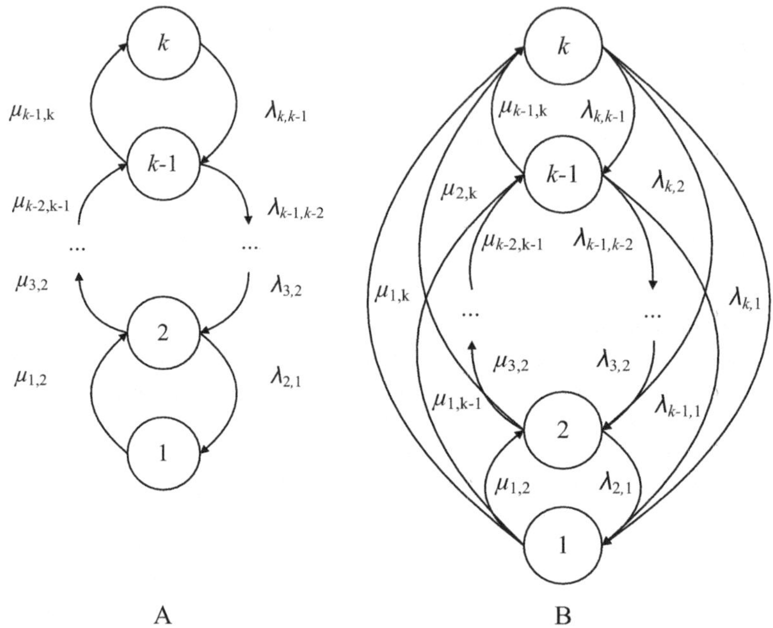

Fig. 2.7 State-transition diagram for repairable element with minor failures and repairs (**a**) and for repairable element with minor and major failures and repairs (**b**)

$$\begin{cases} \dfrac{dp_k(t)}{dt} = \sum\limits_{e=1}^{k-1} \mu_{e,k} p_e(t) - p_k(t) \sum\limits_{e=1}^{k-1} \lambda_{k,e} \\ \dfrac{dp_i(t)}{dt} = \sum\limits_{e=i+1}^{k} \lambda_{e,i} p_e(t) + \sum\limits_{e=1}^{i-1} \mu_{e,i} p_e(t) - p_i(t) \left(\sum\limits_{e=1}^{i-1} \lambda_{i,e} + \sum\limits_{e=i+1}^{k} \mu_{i,e} \right), \\ \qquad i = 2, 3, \ldots, k-1 \\ \dfrac{dp_1(t)}{dt} = \sum\limits_{e=2}^{k} \lambda_{e,1} p_e(t) - p_1(t) \sum\limits_{e=2}^{k} \mu_{1,e} \end{cases} \tag{2.66}$$

with the initial conditions (2.56). Solving this system one obtains the state probabilities $p_i(t)$, $i = 1, \ldots, k$.

When $F(g_i, w) = g_i - w$ for the constant demand level $g_i < w \le g_{i+1}$ the acceptable states where the element performance is above level g_i are $i + 1, \ldots, k$. Therefore, the instantaneous availability is

$$A_i(t) = \sum_{e=i+1}^{k} p_e(t) \tag{2.67}$$

The element's mean instantaneous performance and mean instantaneous performance deficiency can be determined by using (2.63) and (2.64).

In many applications, the steady state or final states probabilities $\lim_{t\to\infty} p_i(t)$ are of interest for the repairable element. As it was said above, if the steady state probabilities exist, the process is called *ergodic*. For the final state probabilities the computations become simpler. The set of differential equations (2.66) is reduced to a set of k algebraic linear equations because for the constant probabilities all time-derivatives are equal to zero, so, $\frac{dp_i(t)}{dt} = 0,\ i = 1, \ldots, k$.

Let the final states probabilities $p_i = \lim_{t\to\infty} p_i(t)$ exist. In order to find the probabilities the following system of algebraic linear equations should be solved

$$\begin{cases} 0 = \sum_{e=1}^{k-1} \mu_{e,k} p_m - p_k \sum_{e=1}^{k-1} \lambda_{k,e} \\ 0 = \sum_{e=i+1}^{k} \lambda_{e,i} p_e + \sum_{e=1}^{i-1} \mu_{e,i} p_e - p_i \left(\sum_{e=1}^{i-1} \lambda_{i,e} + \sum_{e=i+1}^{k} \mu_{i,e} \right),\ i = 2, 3, \ldots, k-1 \\ 0 = \sum_{e=2}^{k} \lambda_{e,1} p_e - p_1 \sum_{e=2}^{k} \mu_{1,e} \end{cases} \tag{2.68}$$

The k equations in (2.68) are not linearly independent (the determinant of the system is zero). An additional independent equation can be provided by the simple fact that the sum of the state probabilities is equal to 1 at any time:

$$\sum_{i=1}^{k} p_i = 1 \tag{2.69}$$

The determination of the reliability function for the repairable multi-state element is based on finding the probability of the event when the element enters the set of unacceptable states the first time. It does not matter which one of the unacceptable states is visited first. It does not matter also how the element behaves after entering the set of unacceptable states the first time.

In order to find the element reliability function $R_i(t)$, for the constant demand w ($g_i < w \le g_{i+1}$), the additional Markov model should be built. All states 1, 2, ..., i of the element corresponding to the performance rates lower than the demand w, should be united in one absorbing state. This absorbing state can be considered now as state 0 and all repairs that return the element from this state back to the set of acceptable states should be forbidden. This corresponds to zeroing all the transition intensities $\mu_{0,m}$ for $m = i + 1$, ..., k. The transition rate $\lambda_{m,0}$ from any acceptable state m ($m > i$) to the united absorbing state 0 is equal to the sum of the transition rates from the state m to all the unacceptable states (states 1, 2, ..., i):

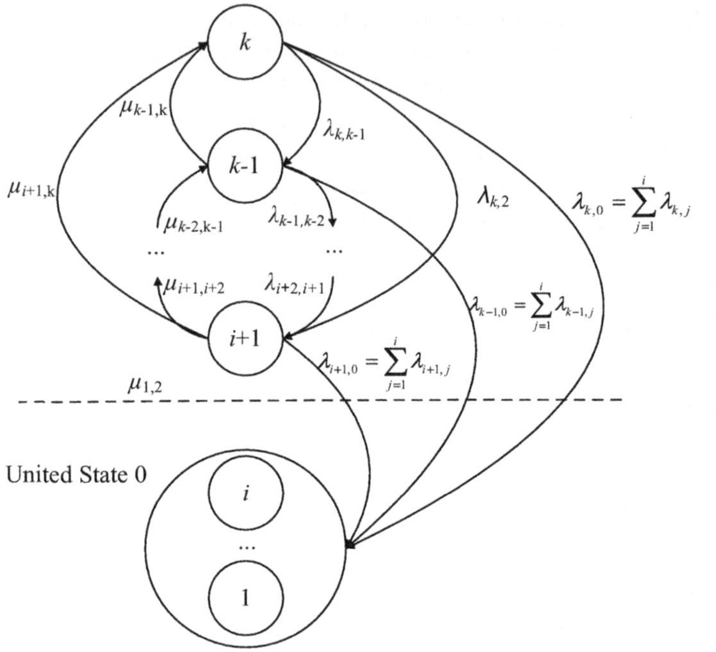

Fig. 2.8 State-transition diagram for determination of reliability function $R_i(t)$ for repairable element (for a constant demand w: $g_i < w < g_i + 1$)

$$\lambda_{m,0} = \sum_{j=1}^{i} \lambda_{m,j}, \; m = k, k-1, \ldots, i+1. \tag{2.70}$$

The state-transition diagram for computation of the reliability function is presented in Fig. 2.8. For this diagram, the state probability $p_0(t)$ characterizes the reliability function of the element because after the first entrance into the absorbing state 0 the element never leaves it: $R_i(t) = 1 - p_0(t)$.

The system of differential equations for determining the reliability function of the element takes the form:

$$\begin{cases} \dfrac{dp_k(t)}{dt} = \sum\limits_{e=i+1}^{k-1} \mu_{e,k} p_e(t) - p_k(t)\left(\sum\limits_{e=i+1}^{k-1} \lambda_{k,e} + \lambda_{k,0}\right) \\ \dfrac{dp_j(t)}{dt} = \sum\limits_{e=j+1}^{k} \lambda_{e,j} p_e(t) + \sum\limits_{e=1}^{j-1} \mu_{e,j} p_e(t) - p_j(t)\left(\sum\limits_{e=i+1}^{j-1} \lambda_{j,e} + \lambda_{j,0} + \sum\limits_{e=j+1}^{k} \mu_{j,e}\right), \\ \quad \text{for } i < j < k \\ \dfrac{dp_0(t)}{dt} = \sum\limits_{e=i+1}^{k} \lambda_{e,0} p_e(t) \end{cases} \tag{2.71}$$

Solving this system under initial conditions

$$p_k(0) = 1,\ p_{k-1}(0) = \cdots = p_i(0) = p_0(0) = 0$$

one obtains the reliability function as

$$R_i(t) = 1 - p_0(t) = \sum_{j=i+1}^{k} p_j(t) \tag{2.72}$$

Obviously, the final state probabilities for the system (2.71) are as follows

$$p_k = p_{k-1} = \cdots = p_{i+1} = 0,\ p_0 = 1,$$

because the element always enters the absorbing state 0 when $t \to \infty$.

Based on computed reliability function $R_i(t) = \sum_{j=i+1}^{k} p_j(t)$ one can find mean time up the first failure, when element performance first time drops under demand level w, where $g_i < w \le g_{i+1}$

$$MTTF_i = \int_0^{\infty} R_i(t)dt \tag{2.73}$$

Once state probabilities, p_i or $p_i(t)$ have been computed, reliability measures are usually obtained based on these probabilities.

Example 2.3 [10] Consider a data processing unit that has $k = 4$ possible performance levels with corresponding task processing speeds: $g_4 = 100\ \text{s}^{-1}$, $g_3 = 80\ \text{s}^{-1}$, $g_2 = 50\ \text{s}^{-1}$ and $g_1 = 0\ \text{s}^{-1}$.

The unit has the following failure rates

$$\lambda_{4,3} = 2\,\text{year}^{-1},\ \lambda_{3,2} = 1\,\text{year}^{-1},\ \lambda_{2,1} = 0.7\,\text{year}^{-1} \quad \text{(for minor failures)},$$

$$\lambda_{3,1} = 0.4\,\text{year}^{-1},\ \lambda_{4,2} = 0.3\,\text{year}^{-1},\ \lambda_{4,1} = 0.1\,\text{year}^{-1} \quad \text{(for major failures)}$$

and the following repair rates

$$\mu_{3,4} = 100\,\text{year}^{-1},\ \mu_{2,3} = 80\,\text{year}^{-1},\ \mu_{1,2} = 50\,\text{year}^{-1} \quad \text{(for minor repairs)},$$
$$\mu_{1,4} = 32\,\text{year}^{-1},\ \mu_{1,3} = 40\,\text{year}^{-1},\ \mu_{2,4} = 45\,\text{year}^{-1} \quad \text{(for major repairs)}.$$

The demand is constant $w = 60\ \text{s}^{-1}$.

Find such element reliability measures as availability, mean performance, mean performance deficiency, reliability function and mean time up to the first failure.

Solution The state-space diagram for the unit is presented in Fig. 2.9a. We assume that initial state is the best state 4.

In order to find the state probabilities the following system of differential equations should be solved:

$$\begin{cases} \dfrac{dp_4(t)}{dt} = -(\lambda_{4,3} + \lambda_{4,2} + \lambda_{4,1})p_4(t) + \mu_{3,4}p_3(t) + \mu_{2,4}p_2(t) + \mu_{1,4}p_1(t) \\ \dfrac{dp_3(t)}{dt} = \lambda_{4,3}p_4(t) - (\lambda_{3,2} + \lambda_{3,1} + \mu_{3,4})p_3(t) + \mu_{1,3}p_1(t) + \mu_{2,3}p_2(t) \\ \dfrac{dp_2(t)}{dt} = \lambda_{4,2}p_4(t) + \lambda_{3,2}p_3(t) - (\lambda_{2,1} + \mu_{2,3} + \mu_{2,4})p_2(t) + \mu_{1,2}p_1(t) \\ \dfrac{dp_1(t)}{dt} = \lambda_{4,1}p_4(t) + \lambda_{3,1}p_3(t) + \lambda_{2,1}p_2(t) - (\mu_{1,2} + \mu_{1,3} + \mu_{1,4})p_1(t) \end{cases}$$

with the initial conditions $p_4(t) = 1$, $p_3(t) = p_2(t) = p_1(t) = 0$.

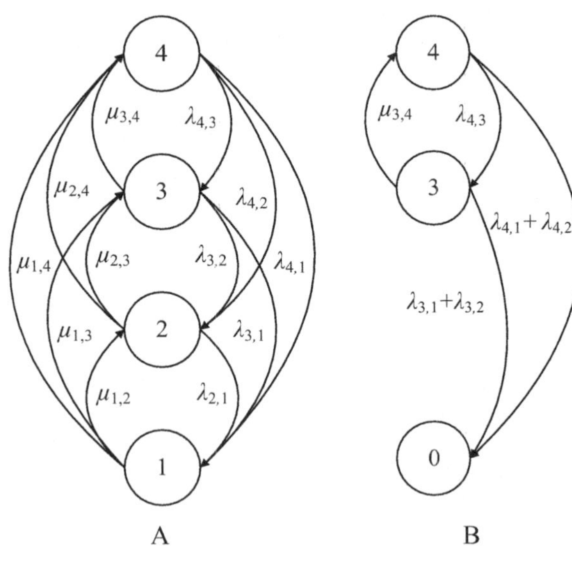

Fig. 2.9 State-transition diagrams for four-state element with minor and major failures and repairs

The element instantaneous availability can be obtained for different constant demand levels

$$A_3(t) = p_4(t), \text{ for } g_3 < w \le g_4,$$
$$A_2(t) = p_4(t) + p_3(t), \text{ for } g_2 < w \le g_3,$$
$$A_1(t) = p_4(t) + p_3(t) + p_2(t) = 1 - p_1(t), \text{ for } g_1 < w \le g_2.$$

These element instantaneous availabilities are presented in Fig. 2.10.

The element mean instantaneous performance at time t is

$$E_t = \sum_{k=1}^{4} g_k p_k(t) = 100p_4(t) + 80p_3(t) + 50p_2(t) + 0p_1(t)$$

For demand $w = 60\ \text{s}^{-1}$ the element availability will be the following

$$A_w(t) = A_2(t).$$

The mean instantaneous performance deficiency (for constant demand $w = 60\ \text{s}^{-1}$) is

$$D_t = \sum_{k=1}^{4} p_k(t)\max(w - g_k, 0) = 10p_2(t) + 60p_1(t).$$

The indices D_t and E_t, as functions of time, are presented in Fig. 2.11.

If one wants to find only the final state probabilities he can do it without solving the system of differential equations. As it was shown above the final state probabilities can be found by solving the system of linear algebraic Eqs. (2.68) in which one of equations is replaced with the Eq. (2.69). In our example, the system of linear

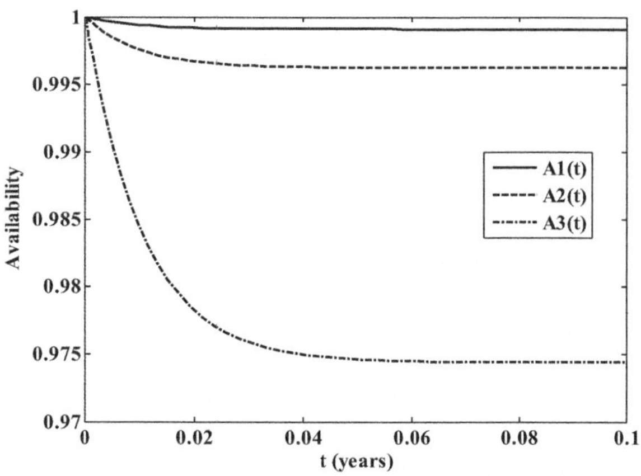

Fig. 2.10 Instantaneous availability of the four-state element

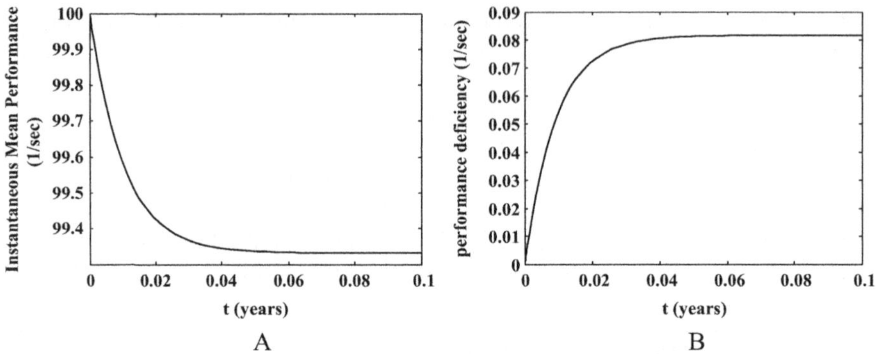

Fig. 2.11 Instantaneous mean performance (**a**) and performance deficiency (**b**) of the four-state element

algebraic equations, which should be solved, takes the form

$$\begin{cases} (\lambda_{4,3} + \lambda_{4,2} + \lambda_{4,1})p_4 = \mu_{3,4}p_3 + \mu_{2,4}p_2 + \mu_{1,4}p_1 \\ (\lambda_{3,2} + \lambda_{3,1} + \mu_{3,4})p_3 = \lambda_{4,3}p_4 + \mu_{2,3}p_2 + \mu_{1,3}p_1 \\ (\lambda_{2,1} + \mu_{2,3} + \mu_{2,4})p_2 = \lambda_{4,2}p_4 + \lambda_{3,2}p_3 + \mu_{1,2}p_1 \\ p_1 + p_2 + p_3 + p_4 = 1 \end{cases}$$

Solving this system, we obtain the final state probabilities:

$$p_1 = \frac{\mu_{1,4}(b_2c_3 - b_3c_2) + \mu_{1,2}(a_2b_3 - a_3b_2) + \mu_{1,3}(a_3c_2 - a_2c_3)}{a_1b_2c_3 + a_2b_3c_1 + a_3b_1c_2 - a_3b_2c_1 - a_1b_3c_2 - a_2b_1c_3},$$

$$p_2 = \frac{\mu_{2,3}(a_1c_3 - a_3c_1) + \mu_{2,4}(b_3c_1 - b_1c_3) + (\lambda_{2,1} + \mu_{2,3} + \mu_{2,4})(a_1b_3 - a_3b_1)}{a_1b_2c_3 + a_2b_3c_1 + a_3b_1c_2 - a_3b_2c_1 - a_1b_3c_2 - a_2b_1c_3},$$

$$p_3 = \frac{\lambda_{3,2}(a_1b_2 - a_2b_1) + (\lambda_{3,2} + \lambda_{3,1} + \mu_{3,4})(a_1c_2 - a_2c_1) + \mu_{3,4}(b_1c_2 - b_2c_1)}{a_1b_2c_3 + a_2b_3c_1 + a_3b_1c_2 - a_3b_2c_1 - a_1b_3c_2 - a_2b_1c_3},$$

$$p_4 = 1 - p_1 - p_2 - p_3,$$

where

$$a_1 = \mu_{1,4} - \mu_{2,4},\ a_2 = \mu_{1,4} - \mu_{3,4},\ a_3 = \mu_{1,4} + \lambda_{4,3} + \lambda_{4,2} + \lambda_{4,1},$$
$$b_1 = \mu_{1,3} - \mu_{2,3},\ b_2 = \mu_{1,3} + \lambda_{3,2} + \lambda_{3,1} + \mu_{3,4},\ b_3 = \mu_{1,3} - \lambda_{4,3},$$
$$c_1 = \mu_{1,2} + \lambda_{2,1} + \mu_{2,3} + \mu_{2,4},\ c_2 = \mu_{1,2} - \lambda_{3,2},\ c_3 = \mu_{1,2} - \lambda_{4,2}.$$

The steady state availability of the element for constant demand $w = 60\ \text{s}^{-1}$ is

$$A = p_4 + p_3,$$

the mean steady state performance is

$$E_{\infty} = \sum_{k=1}^{4} g_k p_k = 100p_4 + 80p_3 + 50p_2 + 0p_1$$

and the mean steady state performance deficiency is

$$D_{\infty} = \sum_{k=1}^{4} p_k \max(w - g_k, 0) = 10p_2 + 60p_1.$$

As one can see in Figs. 2.10 and 2.11, the steady state values of the state probabilities are achieved during a short time. After 0.07 year, the process becomes stationary. Due to this consideration, only the final solution is important in many practical cases. This is especially so for elements with a relatively long lifetime. This is the case in our example if the element lifetime is at least several years. However, if one deals with highly responsible components and takes into account even small information losses at the beginning of the process, the analysis based on a system of differential equations should be performed.

In order to find the element reliability function $R_w(t)$, for the constant demand $w = 60\ \mathrm{s}^{-1}$ ($g_2 < w \le g_3$), the additional Markov model should be built. The states 1 and 2 corresponding to the performance rates lower than the demand w should be united in one absorbing state. This absorbing state can be considered now as state 0 and all repairs that return the element from this state back to the set of acceptable states should be forbidden. This corresponds to zeroing the transition intensities $\mu_{0,3}$ and $\mu_{0,4}$. The transition rates from the acceptable states 3 and 4 to the united absorbing state 0 are equal to the sum of the corresponding transition rates from these states to the unacceptable states 1 and 2. According to (2.70) we obtain $\lambda_{4,0} = \lambda_{4,1} + \lambda_{4,2}$, $\lambda_{3,0} = \lambda_{3,1} + \lambda_{3,2}$.

The state-space diagram for computation of the reliability function $R_w(t)$ is presented in Fig. 2.9b. For this state-space diagram, the state probability $p_0(t)$ characterizes the reliability function of the element because after the first entrance into the absorbing state 0 the element never leaves it.

The system of differential equations for determining the reliability function of the element takes the form:

$$\begin{cases} \dfrac{dp_4(t)}{dt} = -(\lambda_{4,3} + \lambda_{4,2} + \lambda_{4,1})p_4(t) + \mu_{3,4}p_3(t) \\ \dfrac{dp_3(t)}{dt} = \lambda_{4,3}p_4(t) - (\lambda_{3,2} + \lambda_{3,1} + \mu_{3,4})p_3(t) \\ \dfrac{dp_0(t)}{dt} = (\lambda_{4,1} + \lambda_{4,2})p_4(t) + (\lambda_{3,1} + \lambda_{3,2})p_3(t) \end{cases}$$

Solving this system under initial conditions $p_4(0) = 1$, $p_3(0) = p_0(0) = 0$ we obtain the reliability function as $R_w(t) = 1 - p_0(t)$. This function is presented in Fig. 2.12.

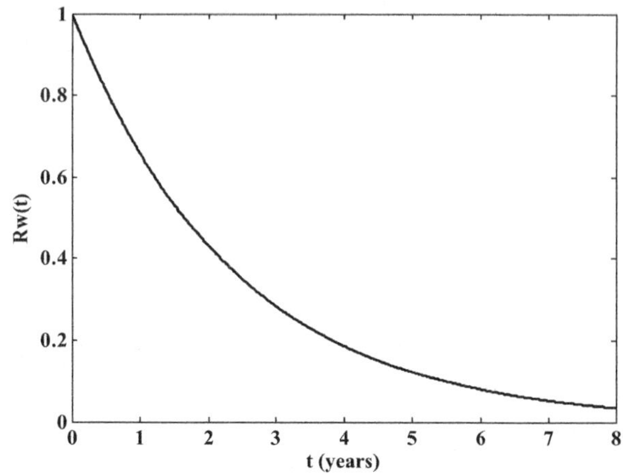

Fig. 2.12 Reliability $R_w(t)$ of the four-state element

When reliability function is known, the mean time up to the first failure (element's capacity dropping below to demand $w = 60\,\text{s}^{-1}$) can be found by using (2.73)

$$MTTF_w = \int_0^\infty R_w(t)dt \approx 2.3 \text{ year}.$$

In order to make the book self-contained, in this chapter we have presented to the reader a basic knowledge about Markov processes. It was considered a methodology of Markov models building for reliability, availability and performability assessment. It will be needed in the next chapters where Markov models should be built for different MSS's elements.

References

1. Bhat U, Miller G (2002) Elements of applied stochastic processes. Wiley, New York
2. Billinton R, Allan R (1996) Reliability evaluation of power systems. Plenum Press, New York
3. Cinlar E (1975) Introduction to stochastic processes. Prentice-Hall, Englewood Cliffs, New York
4. Epstein B, Weissman I (2008) Mathematical models for systems reliability. CRC Press, Taylor & Francis Group, London, New York
5. Gnedenko B, Ushakov I (1995) Probabilistic reliability engineering. Wiley, New York
6. Grimmett G, Stirzaker D (1992) Probability and random processes. Clarendon Press, Oxford
7. International Standard (1995) Application of Markov techniques. International Electrotech Commission IEC 1165
8. Kallenberg O (2002) Foundations of modern probability. Springer, New York, Berlin, Heidelberg
9. Karlin S, Taylor H (1981) A second course in stochastic processes. Academic Press, Orlando, FL

10. Lisnianski A, Frenkel I, Ding Y (2010) Multi-state system reliability analysis and optimization for engineers and industrial managers. Springer, London
11. Nakagawa T (2011) Stochastic processes with applications to reliability theory. Springer, London
12. Papoulis A (1991) Probability, random variables and stochastic processes. McGraw-Hill, NY
13. Ross S (1995) Stochastic processes. Wiley, NY
14. Trivedi K (2002) Probability and statistics with reliability, queuing and computer science applications. Wiley, New York

Chapter 3
L_Z-Transform and Inverse L_Z-Transform of a Discrete-State Continuous-Time Markov Process

During last decades, a specific approach called Universal Generating Function (UGF) technique was widely applied to MSS reliability analysis. The universal generating function technique allows one to find the entire MSS performance distribution based on the performance distributions of its elements by using simple algebraic procedures. This technique is called the method of generalized generating sequences [4]. The basic ideas of the method were primarily introduced by I. Ushakov in the mid-1980s [24, 25]. Then the method was described in the book [18], where one chapter was devoted to UGF. Wide application of the method to MSS reliability analysis began from the mid of 1990s, when the first application was reported [11] and two corresponding papers [7, 16] were published. Since then, the method has been considerably expanded in numerous research papers and in the books [6, 14, 15]. However, the main restriction of this powerful technique is that theoretically, it may be only applied to random variables and, so, concerning MSS reliability, it operates with only steady-state performance distributions. In order to extend the UGF technique application to dynamic MSS reliability analysis where the system and its elements may be characterized by stochastic processes (not only by random variables!) was introduced a special transform for a discrete-states continuous-time Markov process that was called L_Z-transform [10]. In this sense, the L_Z-transform may be considered as UGF technique extension to stochastic processes. Till now it was successfully applied to many real-world problems and important experience has been gathered from industrial settings. These applications will be considered in next chapters of this book.

In this chapter, we shall present mathematical definitions of L_Z-transform and inverse L_Z-transform, its main properties, and numerical examples illustrating its benefits for dynamic MSS reliability assessment. In order to completely understand the material presented in this chapter, the reader should have a basic knowledge about UGF technique. The readers that would like to make their knowledge about this technique deeper can be referred to the basic literature such as [6, 14, 15]. In

A. Lisnianski et al., *Modern Dynamic Reliability Analysis for Multi-state Systems*, Springer Series in Reliability Engineering,
https://doi.org/10.1007/978-3-030-52488-3_3

our book, we shall call UGF technique presented in this literature as a traditional or ordinary UGF technique.

3.1 L_Z-Transform

The ordinary UGF technique allows one to obtain a system's steady-state output performance distribution based on the given steady-state performance distribution of system's elements and the system structure function.

Each multi-state element i, $i = 1, \ldots, m$ in MSS is presented by its steady-state performance as a discrete random variable X_i with probability mass function (*pmf*)

$$\left\{(x_{i1}, p_{i1}), (x_{i2}, p_{i2}), \ldots, \left(x_{in}, p_{in_i}\right)\right\}, \tag{3.1}$$

where x_{ij} is performance of element i at level j and p_{ij} represents probability that element i is staying at level j, $j = 1, \ldots, n_i$.

The system structure function f determines steady-state output performance for the entire MSS as a random variable

$$Y = f(X_1, X_2, \ldots X_m), \tag{3.2}$$

According to the UGF method for each element i that is represented by random variable X_i with *pmf* (3.1) its individual universal generating function is written in the following form

$$u_{X_i}(z) = \sum_{j=1}^{n_i} p_{ij} z^{x_{ij}}, \; i = 1, \ldots, m., \tag{3.3}$$

Then by using Ushakov's universal generating operator Ω_f, which produces UGF for the resulting random variable $Y = f(X_1, X_2, \ldots X_m)$, one obtains the resulting UGF for MSS steady-state output performance distribution based on individual UGF's of the MSS's elements

$$U_Y(z) = \Omega_f\left\{u_{X_1}(z), \ldots, u_{X_m}(z)\right\}, \tag{3.4}$$

The technique for application operator Ω_f is well established for various system structure functions [6, 14, 15]. Based on this technique after some algebra, one can obtain the resulting UGF for random variable Y, which represents the entire MSS's steady-state output performance

$$U_Y(z) = \sum_{k=1}^{K} p_k z^{y_k}, \tag{3.5}$$

where y_k is MSS's steady-state output performance at level k and p_k represents probability that the MSS in steady-state is staying at level k, $k = 1, \ldots, K$.

Based on one-to-one correspondence between discrete random variable and its UGF or, in other words, based on the property of moment generating function *uniqueness* for each discrete random variable (see, for example, [3, 19]) one obtain *pmf* $\{(y_1, p_1), (y_2, p_2), \ldots, (y_n, y_{n_i})\}$ of output random variable Y, when its UGF $U_y(z)$ (expression (3.5)) is known.

Based on steady-state output performance distribution many MSS's reliability measures can be easily found [15, 21].

A main restriction of the UGF technique's application to real-world MSS reliability analysis is caused by the fact that universal generating function was defined only for **random variables**. As one can see, this fact causes to consider performance of each MSS's element as a random variable with *pmf* (3.1), in spite of the fact that in reality it is a discrete-state continuous-time stochastic process [14, 17]. In practice, this important restriction leads to consider only steady-state parameters of MSS. Therefore, now by using ordinary (basic) UGF technique it is impossible to analyze transient modes in MSS, aging MSS, MSS under increasing or decreasing stochastic demand and so on. In the past, some efforts were performed in order to remove this restriction [8, 9, 12]. The method suggested in these works was named as a combined method of stochastic process and UGF technique. Then in [14] was introduced UGF associated with stochastic process. In these works, UGF for discrete-state continuous-time stochastic process was intuitively defined and existence and uniqueness of UGF were only suggested and were not mathematically proven. However, after these works it became clear that such kind of transform may be very useful and essentially extends an area of solving problems. In this case it is possible to work with generic MSS model, which includes the performance stochastic processes $X_i(t)$, $i = 1, 2, \ldots, m$ for each system element i and the system structure function that produces the resulting stochastic process corresponding to the output performance of the entire MSS $Y(t) = f(X_1(t), X_2(t), \ldots X_m(t))$.

In order to remove this essential restriction a special transform should be mathematically defined (introduced) for discrete-state continuous-time stochastic process. This transform should be similar to UGF for discrete random variable in sense that Ushakov's operator Ω_f can be applied. It should be the transform that can be applied to discrete-state continuous-time stochastic process and its important properties such as existence and uniqueness should be proven. Below we consider such transform (named as L_Z-transform). It was introduced in [10]. In this Chapter we consider L_Z-transform existence and uniqueness, study its main properties and demonstrate benefits of its application to MSS reliability analysis by using some numerical examples.

At this stage, only Markov stochastic process will be considered.

3.1.1 L_Z-Transform Definition

We consider a discrete-state continuous-time (DSCT) Markov process [22, 23] $X(t) \in \{x_1, \ldots, x_K\}$, $t \geq 0$, which has K possible states i, ($i = 1, \ldots, K$) where performance level associated with any state i is x_i. This Markov process is completely defined by set of possible states $\mathbf{x} = \{x_1, \ldots, x_K\}$, transitions intensities matrix $\mathbf{A} = \left\| a_{ij}(t) \right\|$, $i, j = 1, \ldots, K$ and by initial states probability distribution that can be presented by corresponding set

$$\mathbf{p}_0 = \left[p_{10} = \Pr\{X(0) = x_1\}, \ldots, p_{K0} = \Pr\{X(0) = x_K\} \right].$$

From now on, we shall use for such Markov process the following notation by using triplet:

$$X(t) = \{\mathbf{x}, \mathbf{A}, \mathbf{p}_0\}, \tag{3.6}$$

If functions $a_{ij}(t) = a_{ij}$ are constants, then the DSCT Markov process is said to be *time-homogeneous*. When $a_{ij}(t)$ are time dependent, then the resulting Markov process is *non-homogeneous*.

Definition L_Z-transform of a discrete-state continuous-time Markov process $X(t) = \{\mathbf{x}, \mathbf{A}, \mathbf{p}_0\}$ is a function $\mathrm{u}(z, t, \mathbf{p}_0)$ defined as

$$L_Z\{X(t)\} = \mathrm{u}(z, t, \mathbf{p}_0) = \sum_{i=1}^{K} p_i(t) z^{x_i}, \tag{3.7}$$

where $p_i(t)$ is a probability that the process is in state i at time instant $t \geq 0$ for any given initial states probability distribution $\mathbf{p}_0$, and z in general case is a complex variable.

As one can see, in (3.7) is using designation $\mathrm{u}(z, t, \mathbf{p}_0)$ that is similar (but not the same!) to ordinary universal generating function designation. It emphasizes the fact that L_Z-transform is genetically tied to its "ancestor"—ordinary universal generating function (UGF). But here should be noticed, that L_Z-transform is not UGF—it is a new mathematical object. As can be seen, L_Z-transform essentially depends on time t and on initial conditions of Markov process $\mathbf{p_0}$.

Example 3.1 Consider a simplest element which has only two states 1 and 2 with corresponding performance levels $x_1 = 0$ and $x_2 = x_{nom}$ respectively. It means that state 1 is a complete failure state and state 2 is a state with nominal performance. The element's failure rate is λ and repair rate is μ.

Suppose that at time instant $t = 0$ the element is in the state 2, so that initial states probability distribution is the following $\mathbf{p}_0 = \{p_{10}, p_{20}\} = \{p_1(0), p_2(0)\} = \{0, 1\}$.

Let's define L_Z-transform for Markov process $X(t)$ that describes the element's stochastic behavior.

Solution The Markov process $X(t)$ for our example is defined by the triplet $X(t) = \{\mathbf{x}, \mathbf{A}, \mathbf{p}_0\}$, where $\mathbf{x}, \boldsymbol{A}, \mathbf{p}_0$ are defined by the following:

- set of possible states $\mathbf{x} = \{x_1, x_2\} = \{0, x_{nom}\}$;
- transitions intensities matrix $\mathbf{A} = (a_{ij}) = \begin{pmatrix} -\mu & \mu \\ \lambda & -\lambda \end{pmatrix}$, $i, j = 1, 2$;
- initial states probability distribution $\mathbf{p}_0 = \{p_{10}, p_{20}\} = \{0, 1\}$.

Therefore, states probabilities of the process $X(t)$ at any time instant $t \geq 0$ will be defined as a solution of the following system of differential equations

$$\begin{cases} \frac{dp_1(t)}{dt} = -\mu p_1(t) + \lambda p_2(t), \\ \frac{dp_2(t)}{dt} = \mu p_1(t) - \lambda p_2(t). \end{cases}$$

under initial conditions $p_1(0) = p_{10} = 0;\ p_2(0) = p_{20} = 1$.

After solving the system, one obtains

$$p_1(t) = \frac{\lambda}{\mu + \lambda} - \frac{\lambda}{\mu + \lambda} e^{-(\lambda+\mu)t},$$

$$p_2(t) = \frac{\mu}{\mu + \lambda} + \frac{\lambda}{\mu + \lambda} e^{-(\lambda+\mu)t}.$$

Therefore, in according to L_Z-transform definition one can obtain L_Z-transform of the given Markov process as follows:

$$L_Z\{X(t)\} = u(z, t, \mathbf{p}_0) = \sum_{i=1}^{2} p_i(t) z^{x_i}$$

$$= \left[\frac{\lambda}{\mu + \lambda} - \frac{\lambda}{\mu + \lambda} e^{-(\lambda+\mu)t}\right] z^0 + \left[\frac{\mu}{\mu + \lambda} + \frac{\lambda}{\mu + \lambda} e^{-(\lambda+\mu)t}\right] z^{x_{nom}}.$$

3.1.2 Existence and Uniqueness

Each discrete-state continuous-time Markov process under certain initial conditions $\mathbf{p}_0$ has only one (unique) L_Z-transform $L_Z\{X(t)\} = u(z, t, \mathbf{p}_0)$ and each L_Z-transform $L_Z\{X(t)\} = u(z, t, \mathbf{p}_0)$ will have only one corresponding DSCT Markov process $X(t)$ developing from these initial conditions.

We'll formulate this as an ***existence and uniqueness property*** of L_Z-transform.

Proposition *Each Discrete-state Continuous-time Markov Process $X(t)$ under certain initial conditions $\mathbf{p}_0$ has one and only one L_Z-transform: $L_Z\{X(t)\} = u(z, t, \mathbf{p}_0)$*

Proof The proof is based on Picard theorem [1]. In the theory of differential equations, the Picard theorem is an important theorem on existence and uniqueness of solutions to system of differential equations with a given initial value problems.

In our case for discrete-state continuous-time Markov process $X(t) = \{\mathbf{x}, \mathbf{A}, \mathbf{p}_0\}$, states probabilities $p_i(t) = \Pr\{X(t) = x_i\}, i = 1, 2, \ldots, K$ are defined by the solution of the following linear system of K ordinary differential equations

$$\begin{cases} \frac{dp_1(t)}{dt} = a_{11}(t)p_1(t) + a_{12}(t)p_2(t) + \cdots + a_{1K}(t)p_K(t), \\ \frac{dp_2(t)}{dt} = a_{21}(t)p_1(t) + a_{22}(t)p_2(t) + \cdots + a_{2K}(t)p_K(t), \\ \ldots \\ \frac{dp_K(t)}{dt} = a_{K1}(t)p_1(t) + a_{K2}(t)p_2(t) + \cdots + a_{KK}(t)p_K(t) \end{cases}, \tag{3.8}$$

under initial conditions $\mathbf{p}_0 = \{p_{10}, p_{20}, \ldots, p_{K0}\}$, where

$$p_1(t_0) = p_{10}, p_2(t_0) = p_{20}, \ldots, p_K(t_0) = p_{K0}, \tag{3.9}$$

In according to Picard's theorem [2], **if coefficients** $a_{kl}(t)$, $(k, l = 1, 2, \ldots K)$ **are continuous functions of time** t, then the system (3.8) has a unique solution

$$p_1(t), p_2(t), \ldots, p_K(t), \tag{3.10}$$

satisfied to initial conditions (3.9).

So, in according to definition of L_Z-transform (3.7) for discrete-state continuous-time Markov process $X(t) = \{\mathbf{x}, \mathbf{A}, \mathbf{p}_0\}$, we shall have

$$L_Z\{X(t)\} = u(z, t, \mathbf{p}_0) = \sum_{i=1}^{K} p_i(t) z^{x_i}, \tag{3.11}$$

where $p_1(t), p_2(t), \ldots, p_K(t)$ are determined as a unique solution of the system (3.8) under initial conditions (3.9).

Therefore, for discrete state continuous time Markov process $X(t) = \{\mathbf{x}, \mathbf{A}, \mathbf{p}_0\}$, where transition intensities $a_{ij}(t)$ are *continuous functions of time*, exists only one (unique) L_Z-transform.

The inverse statement is also true: if it is known $L_Z\{X(t)\} = u(z, t, \mathbf{p}_0) = \sum_{i=1}^{K} p_i(t) z^{x_i}$, where $p_i(t)$ are defined as a solution of the system (3.8) (where coefficients $a_{kl}(t)$, $k, l = 1, 2, \ldots K$ are continuous functions of time t) under initial conditions (3.9), then exists only unique DSCT Markov process $X(t)$, for which $L_Z\{X(t)\} = u(z, t, \mathbf{p}_0) = \sum_{i=1}^{K} p_i(t) z^{x_i}$.

In reliability interpretation L_Z-transform may be applied to an aging system and to a system at burn-in period as well as to a system with constant failure and repair rates. ***The unique condition that should be fulfilled is a continuity of transitions intensities*** $a_{ij}(t)$.

Here we should note that fortunately the results obtain without using L_Z-transform in [9, 10, 13] are still right, because the condition of continuity of functions $a_{ij}(t)$ was fulfilled in these cases.

3.1.3 L_Z-Transform's Properties

Below we consider some important properties of L_Z-transform.

Property 1 *Multiplying DSCT Markov process on constant value a is equal to multiplying corresponding performance level x_i at each state i on this value.*

$$L_Z\{aX(t)\} = \sum_{i=1}^{K} p_i(t)z^{ax_i}, \tag{3.12}$$

Property 2 *L_Z-transform from a single-valued function $f(G(t), W(t))$ of two independent DSCT Markov processes $G(t)$ and $W(t)$ can be found by applying Ushakov's universal generating operator Ω_f to L_Z-transform from $G(t)$ and $W(t)$ processes over all time points $t \geq 0$*

$$L_Z\{f(G(t), W(t))\} = \Omega_f(L_Z\{G(t)\}, L_Z\{W(t)\}).$$

Proof Consider two independent DSCT Markov processes: $G(t) = \{\mathbf{g}, \mathbf{A}, \mathbf{p}_{g0}\}$ and $W(t) = \{\mathbf{w}, \mathbf{B}, \mathbf{p}_{w0}\}$, where

$$\begin{array}{ll} \mathbf{g} = \{g_1, \ldots, g_k\}; & \mathbf{w} = \{w_1, \ldots, w_m\}; \\ \mathbf{A} = \left\| a_{ij}(t) \right\|, i, j = 1, \ldots, k; & \mathbf{B} = \left\| b_{ij}(t) \right\|, i, j = 1, \ldots, m; \\ \mathbf{p}_{g0} = \left\{ p_{10}^{(g)}, p_{20}^{(g)}, \ldots, p_{k0}^{(g)} \right\}; & \mathbf{p}_{w0} = \left\{ p_{10}^{(w)}, p_{20}^{(w)}, \ldots, p_{m0}^{(w)} \right\}. \end{array}$$

In according to the main definition (3.7), one can write the following L_Z-transforms for $G(t)$ and $W(t)$ respectively:

$$\begin{aligned} L_Z\{G(t)\} &= \sum_{i=1}^{k} p_{gi}(t)z^{g_i}, \\ L_Z\{W(t)\} &= \sum_{i=1}^{m} p_{wi}(t)z^{w_i}. \end{aligned} \tag{3.13}$$

The problem is to find L_Z-transform for the resulting DSCT Markov process $Y(t) = f(G(t), W(t))$ that is the single-valued function of these two independent processes $G(t)$ and $W(t)$

$$L_Z\{Y(t)\} = L_Z\{f(G(t), W(t))\}. \tag{3.14}$$

For any time instant $t_r \geq 0$ a DSCT Markov process $G(t)$ is a discrete random variable $G_r = G(t_r)$ with a corresponding *pmf*

$$\Pr\{G(t_r) = g_1\} = p_{g1}(t_r), \ldots, \Pr\{G(t_r) = g_k\} = p_{gk}(t_r).$$

For DSCT Markov process $G(t) = \{\mathbf{g}, \mathbf{A}, \mathbf{p}_{g0}\}$, states probabilities $p_{gi}(t) = \Pr\{G(t) = g_i\}$, $i = 1, 2, \ldots, K$ are defined for any instant $t \geq 0$ by the solution of the following linear system of ordinary differential equations

$$\begin{cases} \frac{dp_{g1}(t)}{dt} = a_{11}(t)p_{g1}(t) + a_{12}(t)p_{g2}(t) + \cdots + a_{1k}(t)p_{gk}(t), \\ \frac{dp_{g2}(t)}{dt} = a_{21}(t)p_{g1}(t) + a_{22}(t)p_{g2}(t) + \cdots + a_{2k}(t)p_{gk}(t), \\ \ldots \\ \frac{dp_{k}(t)}{dt} = a_{k1}(t)p_{g1}(t) + a_{k2}(t)p_{g2}(t) + \cdots + a_{kk}(t)p_{gk}(t) \end{cases} \tag{3.15}$$

under initial conditions $\mathbf{p}_{g0} = \left\{p_{10}^{(g)}, p_{20}^{(g)}, \ldots, p_{k0}^{(g)}\right\}$,

$$p_{g1}(t_0) = p_{10}^{(g)}, p_{g2}(t_0) = p_{20}^{(g)}, \ldots, p_{gk}(t_0) = p_{k0}^{(g)}. \tag{3.16}$$

Analogously, for any time instant $t_r \geq 0$ a DSCT Markov process $W(t)$ is a discrete random variable $W_r = W(t_r)$ with a corresponding *pmf*

$\Pr\{W(t_r) = w_1\} = p_{w1}(t_r), \ldots, \Pr\{G(t_r) = w_k\} = p_{wk}(t_r)$.

For DSCT Markov process $W(t) = \{\mathbf{w}, \mathbf{A}, \mathbf{p}_{w0}\}$, states probabilities $p_{wi}(t) = \Pr\{W(t) = g_i\}$, $i = 1, 2, \ldots, m$ are defined for any instant $t \geq 0$ by the solution of the following linear system of ordinary differential equations

$$\begin{cases} \frac{dp_{w1}(t)}{dt} = b_{11}(t)p_{w1}(t) + b_{12}(t)p_{w2}(t) + \cdots + b_{1m}(t)p_{wm}(t), \\ \frac{dp_{w2}(t)}{dt} = b_{21}(t)p_{w1}(t) + b_{22}(t)p_{w2}(t) + \cdots + b_{2m}(t)p_{wm}(t), \\ \ldots \\ \frac{dp_{wm}(t)}{dt} = b_{m1}(t)p_{w1}(t) + b_{m2}(t)p_{w2}(t) + \cdots + b_{mm}(t)p_{wm}(t) \end{cases}. \tag{3.17}$$

under initial conditions $\mathbf{p}_{w0} = \left\{p_{10}^{(w)}, p_{20}^{(w)}, \ldots, p_{m0}^{(w)}\right\}$,

$$p_{w1}(t_0) = p_{10}^{(w)}, p_{w2}(t_0) = p_{20}^{(w)}, \ldots, p_{wk}(t_0) = p_{m0}^{(w)}. \tag{3.18}$$

In order to define random variable $Y(t_r) = f(G(t_r), W(t_r))$ at any time instant $t_r \geq 0$, one has to evaluate its *pmf*. In other words, one has to evaluate the vector $\mathbf{y} = \{y_1, y_2, \ldots, y_K\}$ of all of the possible values of the resulting random variable $Y(t_r)$ and the vector $\mathbf{q} = \{q_1, q_2, \ldots, q_K\}$ of probabilities that variable $Y(t_r)$ takes these values.

Each possible value $y_1, y_2, \ldots, y_K$ of resulting random variable $Y_r = Y(t_r)$ corresponds to a combination of the values of variables $G_r = G(t_r)$, $W_r = W(t_r)$, and, since the variables are statistically independent, probability of each value

y_i, $i = 1, \ldots, K$ is equal to a product of probabilities of the corresponding values $G(t_r)$, $W(t_r)$. The *pmf* of resulting random variable $Y(t_r) = f(G(t_r), W(t_r))$ is represented in Table 3.1.

As one can see the total number K of possible values of random variable $Y(t_r) = f(G(t_r), W(t_r))$ is $K = km$.

Therefore, vectors $\mathbf{y} = \{y_1, y_2, \ldots, y_K\}$ and $\mathbf{q} = \{q_1(t_r), q_2(t_r), \ldots, q_K(t_r)\}$ presented in the Table 3.1 completely determine resulting *pmf* of random variable $Y_r = Y(t_r) = f(G(t_r), W(t_r)) = f(G_r, W_r)$ at any time instant $t_r \geq 0$.

Now the resulting L_Z-transform for the random variable $Y(t_r) = f(G(t_r), W(t_r))$ can be written at any time instant $t_r \geq 0$

$$L_Z\{Y_r(t)\} = \sum_{j=1}^{K} q_j(t_r) z^{y_j}. \tag{3.19}$$

For any time instant $t_r \in [0, \infty)$ based on the Table 3.1 one can write the following

Table 3.1 Probability mass function of random variable $Y(t_r) = f(G(t_r), W(t_r))$

Values $\mathbf{y} = \{y_1, y_2, \ldots, y_K\}$	Probabilities $\mathbf{q} = \{q_1(t_r), q_2(t_r), \ldots, q_K(t_r)\}$
$y_1 = f(g_1, w_1\}$	$q_1(t_r) = \Pr\{Y(t_r) = y_1\} = p_{g_1}(t_r) p_{w_1}(t_r)$
...	...
$y_m = f(g_1, w_m\}$	$q_m(t_r) = \Pr\{Y(t_r) = y_m\} = p_{g_1}(t_r) p_{w_m}(t_r)$
$y_{m+1} = f(g_2, w_1)$	$q_{m+1}(t_r) = \Pr\{Y(t_r) = y_{m+1}\} = p_{g_2}(t_r) p_{w_1}(t_r)$
...	...
$y_{2m} = f(g_2, w_m)$	$q_{2m}(t_r) = \Pr\{Y(t_r) = y_{2m}\} = p_{g_2}(t_r) p_{w_m}(t_r)$
$y_{2m+1} = f(g_3, w_1)$	$q_{2m+1}(t_r) = \Pr\{Y(t_r) = y_{2m+1}\} = p_{g_3}(t_r) p_{w_1}(t_r)$
...	...
$y_{2m+m} = f(g_3, w_m)$	$q_{2m+m}(t_r) = \Pr\{Y(t_r) = y_{2m+m}\} = p_{g_3}(t_r) p_{w_m}(t_r)$
...	...
$y_{m(k-1)+1} = f(g_k, w_1)$	$q_{mk-m+1}(t_r) = \Pr\{Y(t_r) = y_{mk-m+1}\} = p_{g_k}(t_r) p_{w_1}(t_r)$
...	...
$y_{mk} = f(g_k, w_m)$	$q_{mk}(t_r) = \Pr\{Y(t_r) = y_{mk}\} = p_{g_k}(t_r) p_{w_m}(t_r)$

$$L_Z\{Y_r(t)\} = \sum_{j=1}^{K} q_j(t_r) z^{y_j} = \sum_{i=1}^{k}\sum_{j=1}^{m} p_{gi}(t_r) p_{wj}(t_r) z^{f(g_i, w_j)}. \tag{3.20}$$

By using Ushakov's universal generating operator (UGO) [13] one can re-write expression (3.20) in the following form:

$$L_Z\{Y_r(t)\} = \sum_{i=1}^{k}\sum_{j=1}^{m} p_{gi}(t_r) p_{wj}(t_r) z^{f(g_i, w_j)} = \Omega_f(L_Z\{G_r(t)\}, L_Z\{W_r(t)\}). \tag{3.21}$$

where Ω_f - is Ushakov's universal generating operator, $L_Z\{G_r(t)\} = \sum_{i=1}^{k} p_{gi}(t_r) z^{g_i}$ and $L_Z\{W_r(t)\} = \sum_{j=1}^{m} p_{wj}(t_r) z^{g_j}$ are L_Z-transform for random variables $G_r = G(t_r)$ and $W_r = W(t_r)$ respectively.

Time instant t_r is an arbitrary point in the interval $[0, \infty)$, therefore, for any time instant $t \geq 0$ expression (3.21) may be re-written as the following

$$L_Z\{Y(t)\} = \sum_{i=1}^{k}\sum_{j=1}^{m} p_{gi}(t) p_{wj}(t) z^{f(g_i, w_j)} = \Omega_f(L_Z\{G(t)\}, L_Z\{W(t)\}). \tag{3.22}$$

Therefore, in order to find L_Z-transform of the resulting DSCT Markov process $Y(t)$, which is the single-valued function $Y(t) = f(G(t), W(t))$ of two independent DSCT Markov processes $G(t)$, $W(t)$, one can apply Ushakov's UGO to L_Z-transform of $G(t)$ and $W(t)$ processes over all time points $t \geq 0$.

From computational point of view instead of summarization in expression (3.22) it is better to use the following matrix notation.

Let's designate a column matrix of state probabilities for process $G(t)$ as $\mathbf{P_g}(t)$, and a row matrix of state probabilities for process $W(t)$ as $\mathbf{P_w}(t)$:

$$\mathbf{P_g}(t) = \begin{vmatrix} p_{g1}(t) \\ p_{g2}(t) \\ \cdots \\ p_{gk}(t) \end{vmatrix}, \quad \mathbf{P_w}(t) = |p_{w1}(t), p_{w2}(t), \ldots, p_{wm}(t)|.$$

Note that a matrix with minimal dimension should be chosen as a column matrix. In our case we have supposed that $k \leq m$ and, so, matrix $\mathbf{P_g}(t)$ was written as a column matrix. If $k > m$, then matrix $\mathbf{P_w}(t)$ should be a column matrix and matrix $\mathbf{P_g}(t)$ should be a row matrix.

Then the matrix of states probabilities for the resulting stochastic process $Y(t)$ or in other words, the matrix $\mathbf{Q}(t)$ that defines all coefficients $q_j(t), j = 1, \ldots mk$ in the resulting L_Z-transform

$$L_Z\{Y(t)\} = \sum_{j=1}^{K} q_j(t) z^{y_j}. \tag{3.23}$$

in expression (3.22) can be obtained as a product of matrices $\mathbf{P_g}(t)$ and $\mathbf{P_w}(t)$

$$\mathbf{Q}(t) = \mathbf{P_g}(t) \cdot \mathbf{P_w}(t) = \begin{vmatrix} p_{g1}(t)p_{w1}(t) & p_{g1}(t)p_{w2}(t) & \cdots & p_{g1}(t)p_{wm}(t) \\ p_{g2}(t)p_{w1}(t) & p_{g2}(t)p_{w2}(t) & \cdots & p_{g2}(t)p_{wm}(t) \\ \cdots & \cdots & \cdots & \cdots \\ p_{gk}(t)p_{w1}(t) & p_{gk}(t)p_{w2}(t) & \cdots & p_{gk}(t)p_{wm}(t) \end{vmatrix} \tag{3.24}$$

and $K = mk$.

The corresponding matrix $\mathbf{Y}$ that defines all powers of z (values $y_j, j = 1,\ldots mk$) in the resulting L_Z-transform will be the following

$$\mathbf{Y} = \begin{vmatrix} f(g_1, w_1) & f(g_1, w_2) & \cdots & f(g_1, w_m) \\ f(g_2, w_1) & f(g_2, w_2) & \cdots & f(g_2, w_m) \\ \cdots & \cdots & \cdots & \cdots \\ f(g_k, w_1) & f(g_k, w_2) & \cdots & f(g_k, w_m) \end{vmatrix} \tag{3.25}$$

Such matrix notation is especially useful when MATLAB is used for computation.

Some interesting modification of the matrix method was suggested in [26]. It was shown how it may be used in order to solve the problem of state exploration in the L_Z-transform method.

Expression (3.22) can be extended to general case where the resulting DSCT Markov process $Y(t)$ is the single-valued function $Y(t) = f(X_1(t), X_2(t), \ldots, X_n(t))$ of n independent DSCT Markov processes $X_1(t), X_2(t), \ldots, X_n(t)$:

$$L_Z\{Y(t)\} = \Omega_f(L_Z\{X_1(t)\}, L_Z\{X_2(t)\}, \ldots, L_Z\{X_n(t)\}). \tag{3.26}$$

For computation in according to expression (3.26) may be used ordinary UGF technique (like-terms collection, recursive procedures etc.) over any time instant $t \geq 0$.

As it follows from the properties 1 and 2 L_Z-transform has no a property of linearity.

At first, $L_2\{aX(t)\} \neq aL_2\{X(t)\}$, because

$$L_2\{aX(t)\} = \sum_{i=1}^{K} p_i(t)z^{ax_i} \neq a\sum_{i=1}^{K} p_i(t)z^{x_i} = aL_2\{X(t)\}.$$

At second, $L_Z\{X_1(t) + X_2(t)\} \neq L_Z\{X_1(t)\} + L_Z\{X_2(t)\}$, because

$$L_Z\{X_1(t) + X_2(t)\} = \sum_{i=1}^{k_1}\sum_{j=1}^{k_2} p_{1i}(t)p_{2j}(t)z^{x_{1i}+x_{2j}}$$

and

$$L_Z\{X_1(t)\} + L_Z\{X_2(t)\} = \sum_{i=1}^{K_1} p_{1i}(t) z^{x_{1i}} + \sum_{i=1}^{K_2} p_{2i}(t) z^{x_{2i}}.$$

3.1.4 L_Z-Transform Application to MSS Reliability Analysis

If L_Z-transform

$$L_Z\{Y(t)\} = \sum_{k=1}^{K} p_i(t) z^{y_k}. \tag{3.27}$$

of the entire MSS's output stochastic process $Y(t) \in \{y_1, \ldots y_K\}$ is known, then important system's reliability measures can be easily found.

The system availability at instant $t \geq 0$

$$A(t) = \sum_{y_i \geq 0} p_i(t) \tag{3.28}$$

In other words, in order to find MSS's instantaneous availability one should summarize all probabilities from terms where powers of z are positive or equal to 0.

The system instantaneous mean expected performance at instant $t \geq 0$

$$E(t) = \sum_{k=1}^{K} p_i(t) y_i \tag{3.29}$$

The system average expected performance for a fixed time interval [0, T]

$$E_T = \frac{1}{T} \int_0^T E(t) dt = \frac{1}{T} \sum_{i=1}^{K} y_i \int_0^T p_i(t) dt \tag{3.30}$$

The system instantaneous performance deficiency

$$D(t) = \sum_{i=1}^{K} p_i(t) \min(y_i, 0) \tag{3.31}$$

The system accumulated performance deficiency for a fixed time interval [0, T]

$$D_f = \int_0^T D(t)dt = \sum_{i=1}^{K} \min(y_i, 0) \int_0^T p_i(t)dt \qquad (3.32)$$

We would like to draw the reader attention to the fact that indices (3.29)–(3.32) are an MSS's performability indices and index (3.28) characterizes an MSS's availability.

In order to illustrate L_Z-transform application to MSS reliability analysis, we consider the following example.

Example 3.2 Consider an aging production system that is characterized by 3 possible productivity levels: complete failure, when the system productivity $g_1 = 0$, reduced productivity $g_2 = 400$ units/year, and nominal productivity $g_3 = 800$ units/year. Therefore, the corresponding set of the system's states performances is the following $\mathbf{g} = \{g_1, g_2, g_3\} = \{0, 400, 800\}$.

The transition intensities for the system are represented by the following matrix

$$\mathbf{A}_g = \begin{pmatrix} -(a_{12}^{(g)} + a_{13}^{(g)}) & a_{12}^{(g)} & a_{13}^{(g)} \\ a_{21}^{(g)} & -(a_{21}^{(g)} + a_{23}^{(g)}) & a_{23}^{(g)} \\ a_{31}^{(g)} & a_{32}^{(g)} & -(a_{31}^{(g)} + a_{32}^{(g)}) \end{pmatrix}$$

The state with nominal productivity is the system's initial state $\mathbf{p}_{g0} = (p_{g_1}(0), p_{g_2}(0), p_{g_3}(0)) = (0, 0, 1)$.

So, the system's stochastic performance $G(t) \in \{g_1, g_2, g_3\}$ is defined by DSCT Markov process $G(t) = < \mathbf{g}, \mathbf{A}, \mathbf{p}_{g0}$.

State-transition diagram for Markov process $G(t)$ is presented in Fig. 3.1a.

The system has to satisfy a stochastic demand that is described by two-state DSCT Markov process with minimum performance level $w_1 = 0$ and maximum performance level $w_2 = 500$ units/year. Therefore, the corresponding set of demand's states performances is the following $\mathbf{w} = \{w_1, w_2\} = \{0, 500\}$.

The transition intensities for the demand are represented by the following matrix

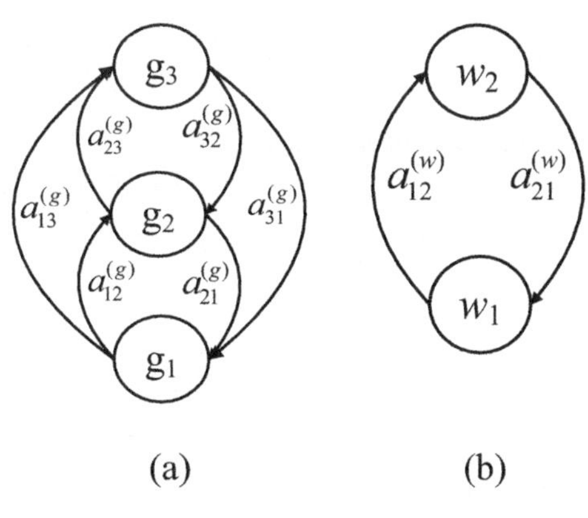

Fig. 3.1 State-transition diagrams for Markov processes $G(t)$ (**a**) and $W(t)$ (**b**)

$$\mathbf{A}_w = \begin{pmatrix} -a_{12}^{(w)} & a_{12}^{(w)} \\ a_{21}^{(w)} & -a_{21}^{(w)} \end{pmatrix}$$

where $a_{12}^{(w)} = 487 + 200\sin(2\pi t)\ \text{year}^{-1}$, $a_{21}^{(w)} = 1095\ \text{year}^{-1}$.

As one can see the demand is seasonally changing—the maximum's duration is increasing at summer and decreasing at winter.

The state with maximum demand is the initial state $\mathbf{p}_{w_0} = \{p_{w_1}(0), p_{w_2}(0)\} = \{0, 1\}$.

The demand's stochastic performance is defined by DSCT Markov process $W(t) = \{\mathbf{w}, \mathbf{A}, \mathbf{p}_{w_0}\}$.

State-transition diagram for Markov process $W(t)$ is presented in Fig. 3.1b.

When the resulting stochastic process $Y(t) = f(G(t), W(t)) = G(t) - W(t)$ falls down to level zero such event is treated as a failure. Processes $G(t),\ W(t)$ are independent.

The problem is to find instantaneous availability and instantaneous mean performance deficiency for this aging multi-state production system under the seasonally changing stochastic demand.

Solution In Fig. 3.2 one can see a block-diagram for the MSS reliability computation. States probabilities $p_{g_i}(t), i = 1, 2, 3$ for Markov processes $G(t)$ can be obtained by the solving of the following system of differential equations:

$$\begin{cases} \frac{dp_{g_1}(t)}{dt} = -(a_{12}^{(g)} + a_{13}^{(g)})p_{g_1}(t) + a_{21}^{(g)}p_{g_2}(t) + a_{31}^{(g)}p_{g_3}(t), \\ \frac{dp_{g_2}(t)}{dt} = a_{12}^{(g)}p_{g_1}(t) - (a_{21}^{(g)} + a_{23}^{(g)})p_{g_2}(t) + a_{32}^{(g)}p_{g_3}(t), \\ \frac{dp_{g_3}(t)}{dt} = a_{13}^{(g)}p_{g_1}(t) + a_{23}^{(g)}p_{g_2}(t) - (a_{31}^{(g)} + a_{32}^{(g)})p_{g_3}(t), \end{cases}$$

under initial conditions $\mathbf{p}_{g_0} = \{p_{g_1}(0), p_{g_2}(0), p_{g_3}(0)\} = \{0, 0, 1\}$.

States probabilities $p_{w_i}(t), i = 1, 2$ for Markov processes $W(t)$ can be obtained by solving of the following system of differential equations:

$$\begin{cases} \frac{dp_{w_1}(t)}{dt} = -a_{12}^{(w)}p_{w_1}(t) + a_{21}^{(w)}p_{w_2}(t), \\ \frac{dp_{w_2}(t)}{dt} = a_{12}^{(w)}p_{w_1}(t) - a_{21}^{(w)}p_{w_2}(t), \end{cases}$$

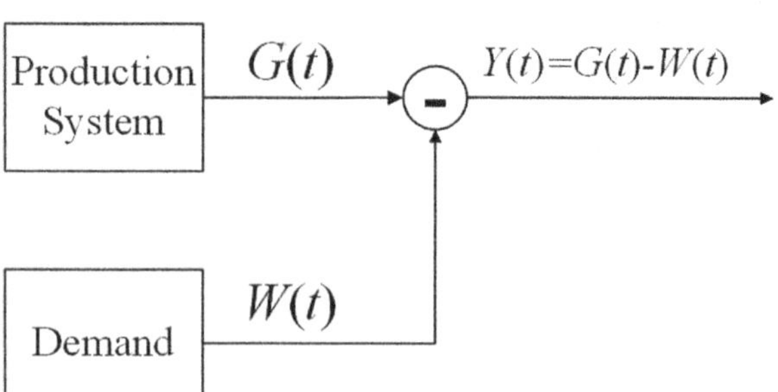

Fig. 3.2 Block-diagram for the MSS reliability computation

under initial conditions $\mathbf{p}_{w_0} = \{p_{w_1}(0), p_{w_2}(0)\} = \{0, 1\}$.

In according to the L_Z–transform definition we obtain

$$L_Z\{G(t)\} = p_{g_1}(t)z^{g_1} + p_{g_2}(t)z^{g_2} + p_{g_3}(t)z^{g_3} = p_{g_1}(t)z^0 + p_{g_2}(t)z^{400} + p_{g_3}(t)z^{800};$$
$$L_Z\{W(t)\} = p_{w_1}(t)z^{w_1} + p_{w_2}(t)z^{w_2} = p_{w_1}(t)z^0 + p_{w_2}(t)z^{500}.$$

Now, to obtain the resulting stochastic process $Y(t) = f\{G(t), W(t)\} = G(t) - W(t)$ we have to calculate the $L_Z\{Y(t)\}$ as follows:

$$\begin{aligned}
L_Z\{Y(t)\} &= L_Z\{G(t) - W(t)\} = \Omega_f(L_Z\{G(t), W(t)\}) \\
&= \Omega_f\left(p_{g_1}(t)z^0 + p_{g_2}(t)z^{400} + p_{g_3}(t)z^{800}, p_{w_1}(t)z^0 + p_{w_2}(t)z^{500}\right) \\
&= \sum_{i=1}^{3}\sum_{j=1}^{2} p_{g_i}(t)p_{w_j}(t)z^{f(g_i, w_j)} = \sum_{i=1}^{3}\sum_{j=1}^{2} p_{g_i}(t)p_{w_j}(t)z^{(g_i - w_j)} \\
&= p_{g_1}(t)p_{w_1}(t)z^{0-0} + p_{g_1}(t)p_{w_2}(t)z^{0-500} + p_{g_2}(t)p_{w_1}(t)z^{400-0} \\
&+ p_{g_2}(t)p_{w_2}(t)z^{400-500} + p_{g_3}(t)p_{w_1}(t)z^{800-0} + p_{g_3}(t)p_{w_2}(t)z^{800-500} \\
&= p_{g_1}(t)p_{w_2}(t)z^{-500} + p_{g_2}(t)p_{w_2}(t)z^{-100} + p_{g_1}(t)p_{w_1}(t)z^0 \\
&+ p_{g_3}(t)p_{w_2}(t)z^{300} + p_{g_2}(t)p_{w_1}(t)z^{400} + p_{g_3}(t)p_{w_1}(t)z^{800}.
\end{aligned}$$

Based on the last expression we obtain:

- MSS's instantaneous availability is calculated in according to expression (3.28) (see Fig. 3.3)

$$A(t) = p_{g_1}(t)p_{w_1}(t) + p_{g_3}(t)p_{w_2}(t) + p_{g_2}(t)p_{w_1}(t) + p_{g_3}(t)p_{w_1}(t)$$

- MSS's mean performance deficiency is calculated in according to expression (3.31) (see Fig. 3.4)

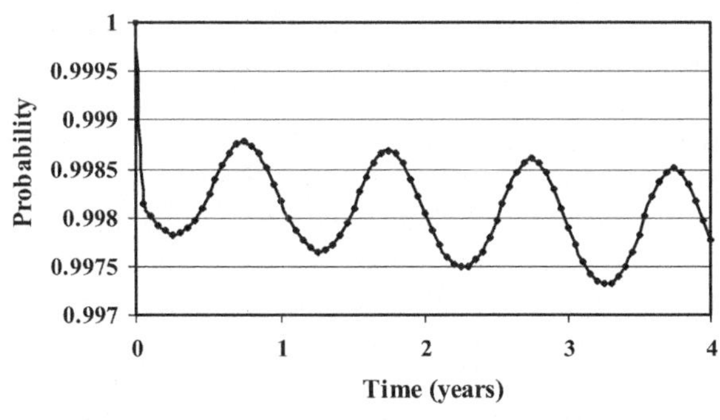

Fig. 3.3 MSS's instantaneous availability

Fig. 3.4 MSS's mean performance deficiency

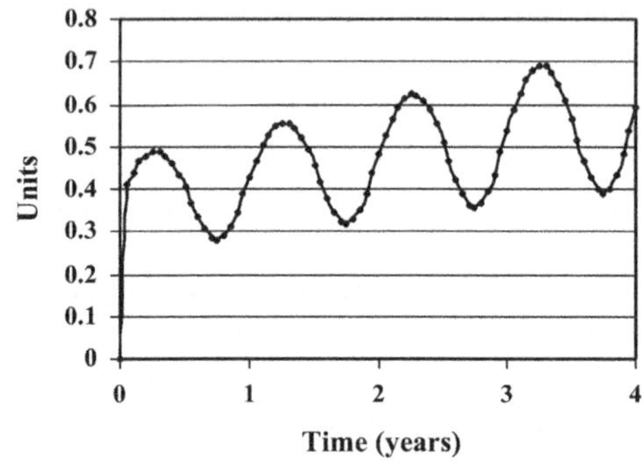

$$D(t) = 500p_{g_1}(t)p_{w_2}(t) + 100p_{g_2}(t)p_{w_2}(t).$$

Note that in order to solve this problem by using straightforward Markov methods one should solve the system of 6 differential equations with variable coefficients.

In general, if one will use a straightforward Markov method for MSS consisting of n different multi-states elements where every element j has k_j different performance levels, he will obtain a model with $K = \prod_{j=1}^{n} k_j$ states.

Therefore, one should solve the system of K differential equations. This number K can be very large even for relatively small MSSs. We shall consider such cases in next chapters of the book, where real world problems will be solved. It will be shown that by using straightforward Markov methods one will be faced with "dimension curse". State-space diagram building or model construction for complex MSSs is a difficult no formalized process that may cause numerous mistakes. In addition, solving models with thousands of states can challenge the available computer resources. If one will use L_Z-transform, he has to build and solve only n separate (relatively simple) Markov models for each element where maximum number of states is defined as $\max\{k_1, \ldots, k_j, \ldots, k_n\}$. All other computations are formalized algebraic procedures. So, by using L_Z-transform method one should solve n systems of $k_1, \ldots, k_n$ differential equations respectively and the total number of differential equations will be $\sum_{i=1}^{n} k_i$. It means that computational complexity decreases drastically when L_Z-transform is used.

In this section was introduced and mathematically defined L_Z-transform for discrete-state continuous-time Markov process, and its main properties were study. It should be emphasized that L_Z-transform is not UGF, it is a new mathematical object.

L_Z-transform application to MSS availability and performability analysis essentially extends a circle of problems that can be solved by using Ushakov's universal generating operator. These are the problems of short-term availability and performability assessment where an impact of initial conditions should be taken into account, the problems of long-term availability and performance assessment for aging systems, systems under stochastic demand and so on.

It was shown that computational complexity is essentially less when L_Z-transform is used compared with traditional Markov method.

3.2 Inverse L_Z-Transform

As a result of L_Z-transform method application to MSS analysis, one obtains L_Z-transform of output performance process for entire system. Based on this such indices as availability, mean expected performance, mean performance deficiency etc., could be found. But often it may be needed to know how much time a system has under specified initial conditions up to its entrance in the failure state, when, for example, a required demand will not be satisfied. Therefore, it is important to know probability distribution of time up to the failure, or reliability function. For this purpose, it is not enough to know L_Z-transform of output performance process. In order to find MSS reliability function $R(t)$ one should completely uncover (reveal) the underlying Markov process when only L_Z-transform of this process is known.

To solve this problem and find the underlying Markov process in [13] an inverse L_Z-transform (designated as L_Z^{-1}-transform) was suggested. When L_Z^{-1}-transform is applied to a given L_Z-transform of resulting output stochastic process, it completely uncovers the underlying Markov process.

Definition Let a function

$$u(z, t, \mathbf{p}_0) = \sum_{i=1}^{K} p_i(t) z^{x_i} \tag{3.33}$$

be L_Z-transform of some unknown discrete-state continuous-time Markov process $X(t) = \{\mathbf{x}, \mathbf{A}, \mathbf{p_0}\}$. Here $p_i(t)$, $i = 1, \ldots, K$ is a probability that Markov process $X(t)$ is in state i at time instant $t \geq 0$, x_i is the performance in this state, $\mathbf{p_0}$ vector of the process states probabilities at initial time instant $t = 0$ and z is a complex variable.

Based on a given L_Z-transform of some DSCT Markov process $X(t)$, inverse L_Z-transform

$$L_Z^{-1}\left\{\sum_{i=1}^{K} p_i(t) z^{x_i}\right\} \tag{3.34}$$

reveals the underlying Markov process $X(t)$. Therefore, the following definition can be written:

$$L_Z^{-1}\{L_Z\{X(t)\}\} = X(t) = \{\mathbf{x}, \mathbf{A}, \mathbf{p_0}\}, \tag{3.35}$$

if all transition intensities in matrix **A** are continuous function of time.

As it was stated above, "to reveal (uncover) underlying Markov process" means to determine for the unknown process:

- a set of possible states **x**,
- a transition intensities matrix **A**,
- a vector of initial conditions $\mathbf{p_0}$.

3.2.1 Computational Procedure for Determining Inverse L_Z-Transform

From computational point of view, the problem of finding L_Z^{-1}-transform can be summarized as the following.

It is given L_Z-transform of some ***unknown*** Markov process $X(t)$

$$L_Z\{X(t)\} = \mathrm{u}(z, t, \mathbf{p}_0) = \sum_{i=1}^{K} p_i(t) z^{x_i}, \tag{3.36}$$

Based on this expression for L_Z-transform one should reveal (uncover) the underlying Markov process $X(t)$, or in other words, to find the set of states **x**, the set of initial conditions $\mathbf{p_0}$ and the matrix $\boldsymbol{A}$ of transitional probabilities of the process.

In reliability interpretation we consider the case when each MSS's component at time $t = 0$ may be in any arbitrary state and any MSS's state with performance lower than any specified demand level is treated as MSS's failure. We suppose that MSS consists of n independent components, where each component i is described by corresponding stochastic process $X_i(t)$. Usually MSS's structure function f, which defines MSS output stochastic process X(t) is known and given by the following expression

$$X(t) = f(X_1(t), X_2(t), \ldots, X_n(t)), \tag{3.37}$$

where $X_i(t) \in \{x_{i1}, x_{i2}, \ldots, x_{ir_i}\}$ is a discrete state continuous time Markov process that describes performance behavior of component i.

Notice that L_Z-transform (3.36) is obtained under specified initial conditions for all system's components. We designate these conditions by using the following notation

$$X_1(0) = x_{1i}, i \in \{1, \ldots r_1\}, \ldots, X_n(0) = x_{nk}, k \in \{1, \ldots, r_n\}, \tag{3.38}$$

where r_m, $m \in \{1, \ldots, n\}$ is a number of performance levels for every component m.

Thus, the problem is to uncover (reveal) the underlying process $X(t)$ based on a given information regarding the system structure function (3.37), L_Z-transform of process $X(t)$ (3.36) and given initial conditions (3.38) for each MSS's component.

3.2.2 Determining Set of States and Set of Initial Conditions for Entire MSS

From expression (3.36), one knows a number of states K of resulting Markov process and corresponding performance in each state i as x_i value. Thus, one can determine set of states $\mathbf{x} = \{x_1, x_2, \ldots, x_K\}$ for underlying process $X(t)$.

Initial conditions for MSS are obtained by given initial conditions of its components (3.38). Thus, initial state of the entire system will be defined by the MSS's structure function (3.37), where corresponding performances of components are determined by their initial states. If, initial conditions of all system's components are given by (3.38), then, in according to MSS's structure function (3.37), initial state of the entire MSS will be the following

$$X(0) = f(x_{1i}, x_{2l}, \ldots, x_{nk}) = x_j,\ j \in \{1, 2, \ldots, K\}. \tag{3.39}$$

Thus, the following initial conditions are determined for entire MSS

$$\mathbf{p_{0j}} = \begin{bmatrix} \Pr\{X(0) = x_1\} = 0, \Pr\{X(0) = x_2\} = 0, \ldots, \Pr\{X(0) = x_j\} \\ = 1, \ldots, \Pr\{X(0) = x_k\} = 0 \end{bmatrix},\ j \in \{1, 2, \ldots, K\} \tag{3.40}$$

It means that at instant $t = 0$ the system is in state j with performance $x_j, j \in [1, \ldots, K]$.

In order to emphasize the fact that L_Z-transform (3.36) is obtained for the given initial states of all MSS's components (3.38) (which then provide initial condition $\mathbf{p_{0j}}$ (3.40) for the entire system), below we will use the following designation for the given L_Z-transform of MSS's resulting (output) performance process $X(t)$:

$$L_Z\{X(t)\} = \sum_{i=1}^{K} p_i^{(j)}(t) z^{x_i},\ j \in \{1, 2, \ldots, K\}. \tag{3.41}$$

3.2.3 Determining Matrix A

The resulting stochastic process $X(t)$ is Markov process that has K states and from expression (3.41), we know all its state probabilities $p_i^{(j)}(t),\ i = 1, \ldots, K$ under the condition that the process begins from state j at time instant $t = 0$.

Generally, probabilities for each of K states can be found by solving the following system of differential equations [1, 22] under given initial conditions $\mathbf{p_{0j}}$ (in matrix notation)

$$\frac{d\mathbf{p}(t)}{dt} = \mathbf{p}(t)\mathbf{A} \tag{3.42}$$

where

$\mathbf{p}(t) = [p_1(t), p_2(t), \ldots, p_K(t)]$ is row-vector of state probabilities;

$\frac{d\mathbf{p}(t)}{dt} = \left[\frac{dp_1(t)}{dt}, \frac{dp_2(t)}{dt}, \ldots, \frac{dp_K(t)}{dt}\right]$ is row-vector of state probabilities' derivatives;

A—transition intensities matrix (that in our case is still unknown),

$$\mathbf{A} = \begin{bmatrix} -(a_{12} + \cdots + a_{1K}) & a_{12} & \cdots & a_{1K} \\ a_{21} & -(a_{21} + \cdots + a_{2K}) & \cdots & a_{2K} \\ \cdots & \cdots & \cdots & \cdots \\ a_{K1} & a_{K2} & \cdots & -(a_{K1} + \cdots + a_{K,K-1}) \end{bmatrix} \quad (3.43)$$

We designate the solution of system (3.42) under initial condition $\mathbf{p_{0\,j}}$ (the system is in state $j, j = 1, 2, \ldots, K$ at instant t = 0) as $\mathbf{p_{Aj}}(t) = \left[p_{A1}^{(j)}(t), p_{A2}^{(j)}(t), \ldots, p_{AK}^{(j)}(t)\right]$.

Ideally, matrix **A** should be determined in order to provide the following equations $p_{A1}^{(j)}(t) = p_1^{(j)}(t), p_{A2}^{(j)}(t) = p_2^{(j)}(t), \ldots, p_{AK}^{(j)}(t) = p_K^{(j)}(t), j = 1, \ldots, K.$

It means that state probabilities of the resulting stochastic process, that can be found as a solution of system (3.42) for initial conditions $\mathbf{p_{0j}}$, and state probabilities from known L_Z-transform of this process (expression (3.36)) should be the same. In practice we should find matrix $\boldsymbol{A}$ in order to provide a minimal difference between all probabilities $p_{Ai}^{(j)}(t)$ and $p_i^{(j)}(t)$, $i = 1, 2, \ldots, K$ for the given $j = 1, \ldots, K$ at any time instant t.

A numerical solution for this problem can be obtained by using Genetic Algorithm (GA). In accordance with GA in each genetic cycle, we randomly generate all $(K-1)\cdot(K-1)$ coefficients a_{ij} in matrix **A** (3.43). We should generate only $(K-1)$ component in each row, because the diagonal component a_{ii} in each row i in matrix **A** is defined as minus the sum of all other components in this row and should not be generated.

Finally, we should find set of coefficients that minimize the accumulated relative error—the measure of difference between solutions of system (3.42) and probabilities, which are presented in L_Z-transform (expression 3.36). This error will be used in GA as a measure of accuracy.

For the given initial conditions $\mathbf{p_{0j}}$, the accumulated relative error, which should be minimized, is the following:

$$ERR_j = \sum_{i=1}^{N_p} w_1 \frac{\left|p_{1A}^{(j)}(t_i) - p_1^{(j)}(t_i)\right|}{p_1^{(j)}(t_i)} + \cdots + \sum_{i=1}^{N_p} w_K \frac{\left|p_{KA}^{(j)}(t_i) - p_K^{(j)}(t_i)\right|}{p_K^{(j)}(t_i)} \quad (3.44)$$

where N_p—number of time points t_i, for which the probability values are compared; $w_1, \ldots, w_k$ are defined as weights of relative error for state 1 and state K, respectively.

The weights of relative errors in some states can be set as small values if their absolute errors are quite small.

In order to determine the number of time points N_p one should analyze functions $p_i^{(j)}(t)$ from the given L_Z-transform. Usually around 1000 time points should be sufficient for the calculation of accumulated relative error (3.44). Approximately a quarter of these points may be taken within a steady state mode and three quarters within transient interval of functions $p_i^{(j)}(t)$.

3.2.4 *The Method Application*

We consider a multi-state system consisting of n multi-state components. Any component j in MSS can have k_j different states corresponding to different performance, represented by the set $\mathbf{g_j} = \{g_{j1}, \ldots, g_{jk_j}\}$, where g_{ji} is the performance rate of component j in the state i, $i \in \{1, 2, \ldots, k_j\}$. The generic MSS model consists of the performance stochastic processes $G_j(t) \in \mathbf{g_j}$, $j = 1, \ldots, n$ for each system component j, and the system structure function that produces the stochastic process corresponding to the output performance of the entire MSS: $G(t) = f(G_1(t), \ldots, G_n(t))$. At first a model of stochastic process should be built for every multi-state component in order to define output performance stochastic process for the entire MSS.

3.2.4.1 Model of Repairable Multi-state Component

Markov performance stochastic process for each component j can be represented by the triplet $G_j(t) = \{\mathbf{g_j}, \mathbf{A_j}, \mathbf{p_{j0}}\}$, where $\mathbf{g_j}$, $\mathbf{A_j}$, $\mathbf{p_{j0}}$ are defined by the following:

- $\mathbf{g_j} = \{g_{j1}, \ldots, g_{jk_j}\}$ - set of possible states;
- $\mathbf{A_j} = \left(a_{lm}^{(j)}(t)\right)$, $l, m = 1, \ldots, k; j = 1, \ldots n$, transition intensities matrix (for aging elements $a_{lm}^{(j)}(t)$ are increasing functions of time);
- $\mathbf{p_{j0}} = \left[p_{10}^{(j)} = \Pr\{G_j(0) = g_{10}\}, \ldots, p_{k_j0}^{(j)} = \Pr\{G_j(0) = g_{k_j0}\}\right]$ initial states probability distribution.

The following system of differential equations can be written for the state probabilities [20].

$$\begin{cases} \frac{dp_{jk_j}(t)}{dt} = \sum\limits_{e=1}^{k_j-1} a_{ek_j}^{(j)}(t)p_{je}(t) - p_{jk}(t) \sum\limits_{e=1}^{k_j-1} a_{k_je}^{(j)}(t) \\ \frac{dp_{ji}(t)}{dt} = \sum\limits_{e=i+1}^{k_j} a_{ei}^{(j)}(t)p_{je}(t) + \sum\limits_{e=1}^{i-1} a_{ei}^{(j)}(t)p_{je}(t) - p_{ji}(t)\left(\sum\limits_{e=1}^{i-1} a_{ie}^{(j)}(t) + \sum\limits_{e=i+1}^{k_j} a_{ie\ i,e}^{(j)}(t)\right) \\ \qquad\qquad\qquad\qquad\qquad\qquad \text{for } 1 < i < k_j \\ \frac{dp_{j1}(t)}{dt} = \sum\limits_{e=2}^{k_j} a_{e1}^{(j)}p_{je}(t) - p_{j1}(t) \sum\limits_{e=2}^{k_j} a_{1e}^{(j)} \end{cases} \tag{3.45}$$

By solving this system (3.45) under initial conditions

$$\mathbf{p}_0^{(j)} = \left[p_{10}^{(j)} = \Pr\{G_j = g_{10}\}, \ldots, p_{k_j 0}^{(j)} = \Pr\{G_j(0) = g_{k_j 0}\}\right] \tag{3.46}$$

one can find (for each element j in the MSS) state probabilities as functions of time:

$$p_{ji}(t) = \Pr\{G_j(t) = g_{ji}\},\ i \in \{1, \ldots, k_j\},\ j \in \{1, \ldots, n\} \tag{3.47}$$

Based on solution (3.45) we can obtain L_Z-transform $L_Z\{G_j(t)\}$ of a Discrete-State Continuous-time (DSCT) Markov process $G_j(t)$

$$L_Z\{G_j(t)\} = \sum_{i=1}^{k_j} p_{ji}(t) z^{g_{ji}} \tag{3.48}$$

where $p_{ji}(t)$ is a probability that the process $G_j(t)$ is in a state with performance g_{ji} at time instant $t \geq 0$ for a given initial states probability distribution $\mathbf{p}_0^{(j)}$, and z in general case is a complex variable. So, for each of the MSS's component the system of differential Eqs. (3.45) should be solved under a given initial conditions (3.46) and a corresponding L_Z-transform (3.48) should be found.

3.2.4.2 Model for Entire Multi-state System

A logical arrangement of the elements in the system is defined by the system structure function $f(G_1(t), \ldots, G_n(t))$. The output performance distribution for the entire MSS *at each time instant t* should be defined based on previously determined states probabilities (3.47) for all elements and A logical arrangement of the elements in the system is defined by the system structure function $f(G_1(t), \ldots, G_n(t))$. The output performance distribution for the entire MSS *at each time instant t* should be defined based on previously determined states probabilities (3.47) for all elements and the system structure function. At this stage L_Z-transform and Ushakov's universal generating operator provide the corresponding computations. L_Z-transform of the output stochastic process for the entire MSS can be defined based on previously determined L_Z-transform for each component j and system structure function f, which produces the output stochastic process of the entire MSS based on stochastic processes of all MSS's elements:

$$G(t) = f(G_1(t), \ldots, G_n(t)) \tag{3.49}$$

In order to find L_Z-transform of the MSS's output performance Markov process $G(t)$, which is the single-valued function $G(t) = f(G_1(t), \ldots, G_n(t))$ of n independent DSCT Markov processes $G_j(t), j = 1, \ldots, n$, one can apply Ushakov's Universal Generating Operator (UGO) [14] to all individual L_Z-transforms $L_Z\{G_j(t)\}$ over all time points $t \geq 0$

$$L_Z\{G(t)\} = \Omega_f(L_Z\{G_1(t)\}, \ldots, L_Z\{G_n(t)\}) = \sum_{i=1}^{K} p_i(t)z^{g_i} \tag{3.50}$$

The technique of Ushakov's operator applying is well established for many different structure functions f.

If all components in some MSS are described by Markov process, the entire MSS is described by Markov process too. So, the resulting process $G(t)$ is Markov process. But after like term collection (summarizing all terms with same powers of z in expression $L_Z\{G(t)\}$), one will have a new expression for L_Z-transform, which is corresponding with new stochastic process G_I(t). This new stochastic process G_I(t) can be considered as the process, which was obtained from the primary Markov process $G(t)$ by lumping (uniting) all states with same performance. In general case lumpability conditions [4, 5, 19] are not fulfilled for this process and the resulting process $G_1(t)$ (after like term collection) is not Markov. This fact does not change the computation of indices in according to expressions (3.28)–(3.32), because all states probabilities for output stochastic process are known. But it will be very important and should be taken into account when reliability function and mean time up to failure will be calculated.

Therefore, after like terms collection in expression (3.50) one will have L_Z-transform for new process $G_1(t)$ with restricted number of states $K_1 < K$, which in general case is not Markov process

$$L_Z\{G_1(t)\} = \sum_{i=1}^{K_1} p_i(t)z^{g_i} \tag{3.51}$$

The possibility of like terms collection is one of the main advantages of UGF and L_Z-transform method, because in many practical cases it helps to restrict drastically a number of states in resulting stochastic process.

In order to use this important advantage and remain in Markov framework we will deal with new Markov process $G_1^M(t)$, which will be equivalent to the process $G_1(t)$ in sense of equality of all probabilities of staying in states with same performances over all time points $t \geq 0$. In other words, L_Z-transform for the process $G_1^M(t)$ is equal to L_Z-transform of the process $G_1(t)$

$$L_Z\left\{G_1^M(t)\right\} = L_Z\{G_1(t)\} = \sum_{i=1}^{K_1} p_i(t)z^{g_i} \tag{3.51}$$

This process $G_1^M(t)$ will be called as *approximated Markov process* for primary output stochastic process $G_1(t)$.

3.2.5 Calculation of Reliability Function and Mean Time to Failure

If L_Z-transform (3.50) of output stochastic process $G_1(t) \in \{g_1, \ldots, g_{K_1}\}$ is known, then reliability function of MSS may be found by the following way.

At first, one has to uncover approximating Markov process $G_1^M(t))$ for output stochastic process $G_1(t)$.

L_Z^{-1}-transform can uncover an underlying Markov process when L_Z-transform of this process is known. Based on the revealed (uncovered) output process MSS reliability function and Mean Time To Failure (MTTF) can be found.

Applying L_Z^{-1}-transform to L_Z-transform in expression (3.51) one can reveal the underlying approximating Markov process $G_1^M(t)$

$$L_Z^{-1}\left\{\sum_{i=1}^{K_1} p_i(t) z^{g_i}\right\} = G_1^M(t) = \{\mathbf{g}, \mathbf{A}, \mathbf{p_0}\}, \tag{3.52}$$

The approximated Markov process has K_1 states $\mathbf{g} = \{g_1 \ldots, g_{K_1}\}$ that are arranged in the ascending order $g_1 \le g_2 \le \cdots \le g_{K_1}$. Reliability function $R(t)$ is treated as probability that the process $G_1^M(t)$ (3.52) that begins at $t = 0$ from state j will not downgrade below specified demand level w_{req} within time t.

In order to find $R(t)$ all states with performance lower than w_{req} should be united in one absorbing state and all transitions from this absorbing state to any other states should be constrained. If $g_k < w_{req}$ and $g_{k+1} \ge w_{req}$, then all states 1, 2,…, k should be united in one absorbing state and all transitions from this absorbing state to any other states should be constrained It means that in matrix $\mathbf{A}$ of the revealed process all elements in rows with numbers equal or lower than k should be zeroed. We designate the matrix as $\mathbf{A_0}$

$$\mathbf{A_0} = \begin{bmatrix} 0 & 0 & \cdots & 0 \\ 0 & 0 & \cdots & 0 \\ \cdots & \cdots & \cdots & \cdots \\ 0 & 0 & \cdots & 0 \\ a_{k+1,1} & a_{k+1,2} & \cdots & a_{k+1,K} \\ \cdots & \cdots & \cdots & \cdots \\ a_{K,1} & a_{K,2} & \cdots & a_{K,K} \end{bmatrix} \tag{3.53}$$

Reliability function $R(t)$ may be found after solving the following system of differential equations in matrix notation

$$\frac{d\mathbf{p}(t)}{dt} = \mathbf{p}(t)\mathbf{A}_0 \tag{3.54}$$

where

- $\mathbf{p}(t) = [p_1(t), p_2(t), \ldots, p_K(t)]$ is row-vector of state probabilities;
- $\frac{d\mathbf{p}(t)}{dt} = \left[\frac{dp_1(t)}{dt}, \frac{dp_2(t)}{dt}, \ldots, \frac{dp_K(t)}{dt}\right]$ is row-vector of state probabilities' derivatives.

Then the reliability function can be obtained

$$R(t) = \sum_{i=k}^{K} p_i(t) \tag{3.55}$$

Based on the reliability function we can obtain the Mean Time to Failure (MTTF) as time up to the first entrance into united absorbing state

$$MTTF = \int_0^\infty R(t)dt \tag{3.56}$$

3.2.6 Numerical Example

Consider a MSS, which consists of three multi-state elements. The MSS's structure, state-transition diagrams of the elements and corresponding performance levels are presented in Fig. 3.5.

Elements' performance levels are as follows:

$$G_1(t) = \{g_{11}, g_{12}\} = \{0, 1.5\},$$
$$G_2(t) = \{g_{21}, g_{22}\} = \{0, 2\},$$

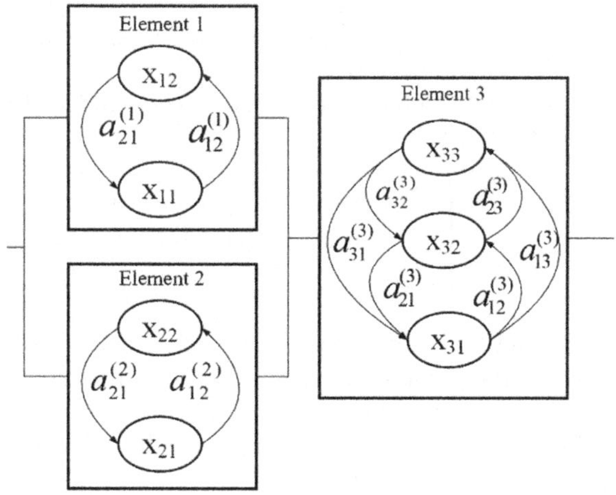

Fig. 3.5 MSS's structure and state-transition diagram of its elements and demand

$$G_3(t) = \{g_{31}, g_{32}, g_{33}\} = \{0, 1.8, 4\}.$$

Failure rates and repair rates are the following:

$a_{21}^{(1)} = 7\ \text{year}^{-1}$	$a_{12}^{(1)} = 100\ \text{year}^{-1}$	$a_{21}^{(2)} = 10\ \text{year}^{-1}$
$a_{12}^{(2)} = 80\ \text{year}^{-1}$,	$a_{12}^{(3)} = 120\ \text{year}^{-1}$,	$a_{21}^{(3)} = 7\ \text{year}^{-1}$,
$a_{32}^{(3)} = 10\ \text{year}^{-1}$,	$a_{23}^{(3)} = 110\ \text{year}^{-1}$.	$a_{13}^{(3)} = 0\ \text{year}^{-1}$.
$a_{31}^{(3)} = 0\ \text{year}^{-1}$.		

The MSS structure function is given by the expression

$$G(t) = f(G_1(t), G_2(t), G_3(t)) = \min\{G_1(t) + G_2(t), G_3(t)\}.$$

The system has to satisfy a constant demand $w_{req} = 1$. The problem is to calculate the system availability and reliability for time period T as well as mean time to failure.

Solution Applying the procedure, described above we proceed as follows. According to the Markov method we build the systems of differential equations for each element (using the state-transitions diagrams presented in Fig. 3.5):

For the first element:

$$\begin{cases} \frac{dp_{11}(t)}{dt} = -a_{12}^{(1)} p_{11}(t) + a_{21}^{(1)} p_{12}(t), \\ \frac{dp_{12}(t)}{dt} = -a_{21}^{(1)} p_{12}(t) + a_{12}^{(1)} p_{11}(t). \end{cases}$$

The initial conditions are: $\mathbf{p_{10}} = \{p_{11}(0), p_{12}(0)\} = \{0, 1\}$.

For the second element:

$$\begin{cases} \frac{dp_{21}(t)}{dt} = -a_{12}^{(2)} p_{21}(t) + a_{21}^{(2)} p_{12}(t), \\ \frac{dp_{22}(t)}{dt} = -a_{21}^{(2)} p_{22}(t) + a_{12}^{(2)} p_{21}(t). \end{cases}$$

The initial conditions: $\mathbf{p_{20}} = \{p_{21}(0), p_{22}(0)\} = \{0, 1\}$.

For the third element:

$$\begin{cases} \frac{dp_{31}(t)}{dt} = -a_{12}^{(3)} p_{31}(t) + a_{21}^{(3)} p_{32}(t), \\ \frac{dp_{32}(t)}{dt} = a_{32}^{(3)} p_{33}(t) - (a_{21}^{(3)} + a_{23}^{(3)}) p_{32}(t) + a_{12}^{(3)} p_{31}(t), \\ \frac{dp_{33}(t)}{dt} = -a_{32}^{(3)} p_{33}(t) + a_{23}^{(3)} p_{32}(t). \end{cases}$$

The initial conditions: $\mathbf{p_{30}} = \{p_{31}(0), p_{32}(0), p_{33}(0)\} = \{0, 0, 1\}$.

By solving these three systems of differential equations, one obtains L_Z-transforms for three processes:

Process $G_1(t)$: $\mathbf{g_1} = \{g_{11}, g_{12}\} = \{0, 1.5\}$, $\mathbf{p_1}(t) = \{p_{11}(t), p_{12}(t)\}$, $\mathbf{p_{10}} = \{p_{11}(0), p_{12}(0)\} = \{0, 1\}$.

The associated L_Z-transform: $L_Z\{G_1(t)\} = \sum_{i=1}^{2} p_{1i}(t) z^{g_{1i}}$.

Process $G_2(t)$: $\mathbf{g}_2 = \{g_{21}, g_{22}\} = \{0, 2.0\}$, $\mathbf{p}_2(t) = \{p_{21}(t), p_{22}(t)\}$,

$$\mathbf{p_{20}} = \{p_{21}(0), p_{22}(0)\} = \{0, 1\}.$$

The associated L_Z-transform: $L_Z\{G_2(t)\} = \sum_{i=1}^{2} p_{2i}(t) z^{g_{2i}}$.

Process $G_3(t)$: $\mathbf{g}_3 = \{g_{31}, g_{32}, g_{33}\} = \{0, 1.8, 4.0\}$, $\mathbf{p}_3(t) = \{p_{31}(t), p_{32}(t), p_{33}(t)\}$, $\mathbf{p_{30}} = \{p_{31}(0), p_{32}(0), p_{33}(0)\} = \{0, 0, 1\}$.

The associated L_Z-transform: $L_Z\{G_3(t)\} = \sum_{i=1}^{3} p_{3i}(t) z^{g_{3i}}$.

Now by using Ushakov's operator Ω_f over all L_Z-transforms of individual elements we can obtain L_Z-transform $L_Z\{G(t)\}$ associated with output performance stochastic process $G(t)$ of the entire MSS:

$L_Z\{G(t)\} = \Omega_f(L_Z\{G_1(t)\}, L_Z\{G_2(t)\}, L_Z\{G_3(t)\})$,

where the system structure function is as follows:

$$G(t) = f(G_1(t), G_2(t), G_3(t)) = \min\{G_1(t) + G_2(t), G_3(t)\}.$$

Based on the known rules for series-parallel MSS (Lisnianski et al. 2010), after like terms collection we finally obtain

$$L_Z\{G(t)\} = \sum_{i=1}^{5} p_i(t) z^{g_i}$$

where

$$\begin{aligned}
&g_1 = 0, \quad p_1(t) = p_{11}(t)p_{21}(t) + p_{31}(t)p_{12(t)} + p_{31}(t)p_{11}(t)p_{22}(t);\\
&g_2 = 1.5,\ p_2(t) = p_{12}(t)p_{21}(t)(p_{32}(t) + p_{33}(t));\\
&g_3 = 1.8,\ p_3(t) = p_{32}(t)p_{22}(t);\\
&g_4 = 2.0,\ p_4(t) = p_{33}(t)p_{11}(t)p_{22}(t);\\
&g_5 = 3.5,\ p_5(t) = p_{33}(t)p_{12}(t)p_{22}(t).
\end{aligned}$$

These two sets $\mathbf{g} = \{g_1, \ldots, g_5\} = \{0, 1.5, 1.8, 2.0, 3.5\}$ and $\mathbf{p}(t) = \{p_1(t), \ldots, p_5(t)\}$ define performance rates and states probabilities of output performance stochastic process $G_1(t)$ for the entire MSS.

Now by using inverse L_Z^{-1}-transform the underlying approximated Markov process $G_1^M(t)$ can be revealed.

As one can see from the obtained L_Z-transform, the underlying output process has 5 states $\mathbf{g} = \{g_1, \ldots, g_5\} = \{0, 1.5, 1.8, 2.0, 3.5\}$. Corresponding states probabilities are such as the follows $\mathbf{p}(t) = \{p_1(t), \ldots, p_5(t)\}$. The initial state is the best state with performance g_5.

In general, states probabilities for 5-states Markov process can be obtained from the following system of ordinary differential equations:

$$\begin{cases} \frac{dp_{1A}(t)}{dt} = -\left(\sum_{i=2}^{5} a_{1i}\right)p_{1A}(t) + \sum_{i=2}^{5} a_{i1}p_{iA}(t), \\ \frac{dp_{2A}(t)}{dt} = -\left(\sum_{i=1, i\neq 2}^{5} a_{2i}\right)p_{2A}(t) + \sum_{i=1, i\neq 2}^{5} a_{i2}p_{iA}(t), \\ \frac{dp_{3A}(t)}{dt} = -\left(\sum_{i=1, i\neq 3}^{5} a_{3i}\right)p_{3A}(t) + \sum_{i=1, i\neq 3}^{5} a_{i3}p_{iA}(t), \\ \frac{dp_{4A}(t)}{dt} = -\left(\sum_{i=1, i\neq 4}^{5} a_{4i}\right)p_{4A}(t) + \sum_{i=1, i\neq 4}^{5} a_{i4}p_{iA}(t), \\ \frac{dp_{5A}(t)}{dt} = -\left(\sum_{i=1}^{4} a_{5i}\right)p_{5A}(t) + \sum_{i=1}^{4} a_{i5}p_{iA}(t). \end{cases}$$

under given initial conditions: $p_{1A}(0) = p_{2A}(0) = p_{3A}(0) = p_{4A}(0) = 0$, $p_{5A}(0) = 1$.

The solution of this system $p_{1A}(t), p_{2A}(t), p_{3A}(t), p_{4A}(t), p_{5A}(t)$ should be numerically close to probabilities $p_1(t), p_2(t), p_3(t), p_4(t), p_5(t)$ that were found above by using L_Z-transform.

In accordance with the GA procedure transition intensities a_{ij} need be found in order to minimize error for 5-states Markov process:

$$Err = \sum_{i=1}^{N_p} \frac{|p_{1A}(t_i) - p_1(t_i)|}{p_1(t_i)} + \cdots + \sum_{i=1}^{N_p} \frac{|p_{5A}(t_i) - p_5(t_i)|}{p_5(t_i)}$$

In each GA cycle in this example the solution $p_{1A}(t), p_{2A}(t), p_{3A}(t), p_{4A}(t), p_{5A}(t)$ should be obtained for the period of 0.15 year. During 0.15 year, the transient mode for the solution $p_1(t), p_2(t), p_3(t), p_4(t), p_5(t)$ will be completely over (finished) and the process will be in steady-state. For the comparison, we shall take 1000 points—one point for 0.00015 year.

The population size in the GA is 100. The offspring will mutate with probability, which avoids premature convergence to a local optimum and facilities jumps in the solution space. The mutation probability is 0.005. The convergence criterion in the proposed GA is set as satisfying both a minimal number of genetic cycles (500 cycles) and a number of genetic cycles without improving the solution performance (50 cycles). The GA converges to optimal solutions by performing about 700 genetic cycles.

So, the underlying Markov process $G_1^M(t)$ was completely revealed:

$$G(t) = \{\mathbf{g}, \mathbf{A}, \mathbf{p_0}\},$$

where

- $\mathbf{g} = \{g_1, \ldots, g_5\} = \{0, 1.5, 1.8, 2.0, 3.5\}$,

- $\mathbf{A} = \begin{bmatrix} -295 & 95 & 120 & 80 & 0 \\ 10.63 & -157.51 & 96.88 & 0 & 50.0 \\ 8.0 & 50.0 & -299.09 & 191.09 & 50.0 \\ 10.07 & 0 & 10.75 & -408.32 & 387.5 \\ 0 & 11.31 & 12.75 & 7 & -31.06 \end{bmatrix}$;
- $\mathbf{p_0} = \{0, 0, 0, 0, 1\}$.

In order to find the reliability function $R(t)$, all transitions from worst state 1 should be constrained. It means that all a_{1i}, $i = 2, \ldots, K$ should be zeroed. Therefore, we will have the matrix

$$\mathbf{A}_0 = \begin{bmatrix} 0 & 0 & 0 & 0 & 0 \\ 10.63 & -157.51 & 96.88 & 0 & 50.0 \\ 8.0 & 50.0 & -299.09 & 191.09 & 50.0 \\ 10.07 & 0 & 10.75 & -408.32 & 387.5 \\ 0 & 11.31 & 12.75 & 7 & -31.06 \end{bmatrix}.$$

Reliability function may be obtained as following:

$$R(t) = \sum_{i=2}^{5} p_i(t),$$

where functions $p_i(t)$ are obtained by solving the system of ordinary differential equations

$$\frac{d\mathbf{p}(t)}{dt} = \mathbf{p}(t)\mathbf{A}_0$$

under initial conditions $\mathbf{p_0} = \{0, 0, 0, 0, 1\}$.

In Fig. 3.6 one can see graphs of reliability functions, calculated by using inverse L_Z-transform and by a conventional straightforward Markov method [14].

As one can see, the reliability curves representing those two solutions are positioned so close together that the difference between them cannot be distinguished visually. (The difference is in the 4th digit after the decimal point).

Now mean time to failure $MTTF_L$ can be obtained

$$MTTF_L = \int_0^{\infty} R_L(t)dt = 0.569\text{years}.$$

Notice that mean time to failure $MTTF_M$ calculated by using straightforward Markov method is almost the same

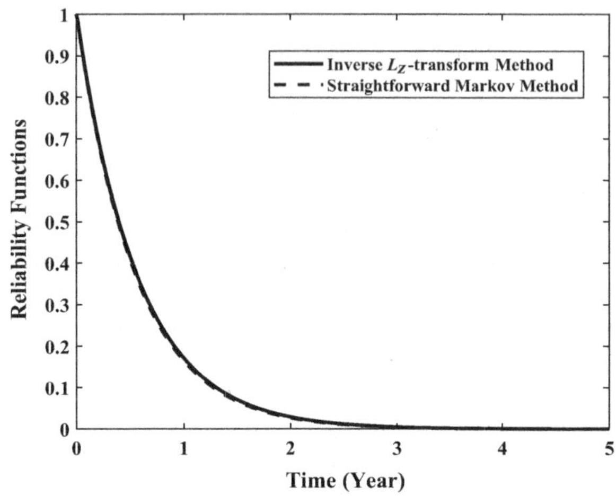

Fig. 3.6 Graphs of reliability functions calculated by using straightforward Markov method and by using inverse L_Z-transform

$$MTTF_M = \int_0^\infty R_M(t)dt = 0.568\,\text{years}.$$

The error for $MTTF_L$ calculation is less than 0.2% which is a very good accuracy for reliability problems where high uncertainty in failure data is expected.

It should be noticed that resulting output stochastic process that was found by using L_Z-transform after like term collection and approximating Markov process that was revealed by inverse L_Z-transform has only 5 states. Markov process, which was built by using straightforward Markov method in [14] in order to perform reliability analysis for this example has 12 states. Therefore, even for this simple example the computational complexity decreases drastically with the use of suggested method.

3.3 Summary

In the chapter was presented approach for computation reliability, availability and performability of MSS in dynamic modes. The approach application significantly simplified a computation burden in comparison with traditional straightforward methods.

Generally, the suggested approach can be presented by the following steps.

1. Solving ordinary differential equations for Markov model of each element in order to obtain state probabilities as functions of time and determine L_Z-transform for each individual element.
2. Obtaining resulting L_Z-transform for entire MSS output performance Markov process $G(t)$ by using Ushakov's universal generating operator and corresponding techniques. After like term collection L_Z-transform for resulting non-Markov

process $G_{-1}(t)$ will be obtained. Based on this L_Z-transform availability and performability indices for entire MSS can be easily obtained.

3. To evaluate MSS reliability indices, inverse L_Z-transform should be used in order to uncover approximating Markov process $G_1^M(t)$ for non-Markov process G_1-(t).
4. Investigating uncovered Markov process $G_1^M(t)$ (that is substantially simplier than process $G(t)$) in order to obtain reliability function and mean time to failure.

L_Z-transform and inverse L_Z-transform can be applied to dynamic reliability, availability and performability analysis of MSS in transient modes, for short-term analysis, for aging MSS and MSS under stochastic demand.

The first application of L_Z-transform was demonstrated in [20] where by using the method an optimal age replacement policy for MSS was determined.

In the next chapters we shall consider applications of this approach to real-world problems. Now it was accumulated enough experience in such applications. Because of the restricted volume we cannot present in the book all these applications. Here we only note that some additional interesting applications of the method one can find in [1].

References

1. Bolvashenkov I, Herzog H-G, Frenkel I, Khvatskin L, Lisnianski A (2018). Safety-critical electrical drives. Topologies, reliability and performance. Springer, Cham, Switzerland
2. Coddington E, Levinson N (1955) Theory of ordinary differential equations. McGraw-Hill, NY
3. Feller W (1970) An introduction to probability theory and its applications. Wiley, NY
4. Gnedenko B, Ushakov I (1995) Probabilistic reliability engineering. Wiley, NY
5. Kemeny J, Snell J (1960) Finite Markov chains. Van Nostrand, NY
6. Levitin G (2005) Universal generating function in reliability analysis and optimization. Springer, London
7. Levitin G, Lisnianski A, Ben-Haim H et al (1998) Redundancy optimization for series-parallel multi-state systems. IEEE Trans Reliab 47:165–172
8. Lisnianski A (2004) Universal generating function technique and random process methods for multi-state system reliability analysis. In: Proceedings of the 2nd international workshop in applied probability (IWAP2004). Piraeus, Greece, pp 237–242
9. Lisnianski A (2007) Extended block diagram method for a multi-state system reliability assessment. Reliab Eng Syst Saf 92(12):1601–1607
10. Lisnianski A (2012) L_Z-transform for a discrete-state continuous-time Markov process and its applications to multi-state system reliability. In: Lisnianski A, Frenkel I (eds) Recent advances in system reliability. Signatures, multi-state systems and statistical inference. Springer, London, pp 79–96
11. Lisnianski A, Ben Haim H, Elmakis D (1994) Redundancy optimization for power station. In: Proceedings of the 10th international conference of the israel society for quality, Jerusalem, 14–17 Nov 1994, pp 313–319
12. Lisnianski A, Ding Y (2009) Redundancy analysis for repairable multi-state system by using combined stochastic process methods and universal generating function technique. Reliab Eng Syst Saf 94:1788–1795

13. Lisnianski A, Ding Y (2014) Inverse L_Z-transform for a discrete-state continuous-time Markov process and its application to multi-state system reliability analysis. In: Frenkel I, Lisniansk A, Karagrigoriou A, Kleiner A (eds) Applied reliability and risk analysis: probabilistic models and statistical inference. Wiley, New York, pp 43–58
14. Lisnianski A, Frenkel I, Ding Y (2010) Multi-state system analysis and optimization for engineers and industrial managers. Springer, London
15. Lisnianski A, Levitin G (2003) Multi-state system reliability: assessment, optimization and applications. World Scientific, Singapore
16. Lisnianski A, Levitin G, Ben Haim H et al (1996) Power system structure optimization subject to reliability constraints. Electric Power Syst Res 39:145–152
17. Natvig B (2011) Multistate systems reliability. Theory with Applications, Wiley, NY
18. Reinshke K, Ushakov I (1988) Application of graph theory for reliability analysis. Radio I Sviaz, Moscow (in Russian and German)
19. Ross S (2000) Introduction to probability models. Academic, Boston
20. Sheu S-H, Zhang Z (2013) An optimal age replacement policy for multi-state systems. IEEE Trans Reliab 62(3):722–735
21. Singh C, Billinton R (1975) Frequency and duration concepts in system reliability evaluation. IEEE Trans Reliab R 24(1):31–36
22. Trivedi K (2002) Probability and statistics with reliability, queuing and computer science applications. Wiley, NY
23. Trivedy K, Bobbio A (2017) Reliability and availability engineering. Modeling, analysis, and applications. Cambridge University Press, UK
24. Ushakov I (1986) A universal generating function. Sov J Comput Syst Sci 24:37–49
25. Ushakov I (1987) Optimal standby problem and a universal generating function. Sov J Comput Syst Sci 25:61–73
26. Yi H, Cui L (2017) A study on repairable series systems with Markov repairable units. In: Frenkel I, Karagrigoriou A, Lisnianski A (eds) Recent advances in multi-state systems reliability: theory and applications. Springer, Cham, Switzerland

Chapter 4
Short-Term Availability, Performability and Reliability Analysis for Power Systems

In this chapter we present applications of L_Z-transform method to power system short-term analysis. Below will be considered availability and performability assessment for coal fired power units as well as for combine cycle units. By using inverse L_Z-transform for coal fired power plant will be also presented short-term risk function and reliability evaluation and estimating of mean time to failure. It will be shown how these parameters may be used in power system dispatch for important operating decisions making in real time.

4.1 Short-Term Availability and Performability Evaluation for Coal Fired Power Station by Using L_Z-Transform

In this section, we consider a short-term availability and performability evaluation for coal-fired power station where each power-generating unit is presented by multi-state Markov model. By using L_Z-transform one can drastically simplify computation. The proposed approach is very useful for power system short-term analysis and for making short-term operating decisions. In order to illustrate the technique in the section presented short-term availability and performability evaluation for power station with different coal fired generating units.

4.1.1 Multi-state Model for a Single Coal Fired Power Generating Unit

Multi-state models are widely used in the field of power system reliability assessment [1]. It has been recognized in [3], that using simple two-state models for large

A. Lisnianski et al., *Modern Dynamic Reliability Analysis for Multi-state Systems*, Springer Series in Reliability Engineering,
https://doi.org/10.1007/978-3-030-52488-3_4

generating units in generating capacity adequacy assessment can yield pessimistic appraisals. In order to more accurately assess power system reliability, many utilities now use multi-state models instead of two-state representations [13]. A technique called the apportioning method [3] is usually used to create steady-state multi-state generating unit models, based on real world statistical data for generating units. Using this technique, steady-state probabilities of units residing at different generating capacity levels can be defined. When the short-term behavior of an MSS is studied, the investigation cannot be based on steady-state (long-term) probabilities. This investigation should use the MSS model where transition intensities between any states of the model are known.

One type of such model was suggested in [6]. In this study, transition intensities were defined for a simplified multi-state Markov model where transitions to each derated state were possible from only one state with a nominal generating capacity. In practice, transitions are possible between all states of the model. Such model was presented in [10], where general multi-state Markov model was considered for coal fired generating unit. In this paper was suggested the method for transition intensities estimation from actual unit failures (deratings) and repairs statistics, which is presented by the observed realization of generating capacity stochastic process. Based on this a four-state Markov model was built for the coal fired unit with nominal generating capacity 575 MW. It is presented in Fig. 4.1 by using graph with corresponding states and transitions.

P_1, P_2, P_3, P_4—steady-state probabilities of corresponding states 1, 2, 3, 4.

For the considered generating unit the transition intensities matrix (4.1) was computed in [12]:

$$A = \begin{vmatrix} -0.0933 & 0.0800 & 0.0133 & 0 \\ 0.0294 & -0.3823 & 0.3235 & 0.0294 \\ 0 & 0.0288 & -0.3846 & 0.3558 \\ 0.0002 & 0.0001 & 0.0007 & -0.0010 \end{vmatrix} \tag{4.1}$$

Each element a_{ij} in matrix (4.1) is represented by such units as 1/h.

In general case multi-state Markov model may consist of m states.

It was shown that such important indices as Loss Of Load Probability (LOLP), Expected Energy Not Supplied to consumers (EENS), etc., which were found for the unit for a short time, are essentially different from those found by a long-term evaluation.

Usually in each power station there are a number of generating units. It is naturally to expect that short-term availability and performability indices for entire power station also may be different from long-term indices. In order to calculate these short-term indices by using straight forward Markov method one should built Markov model for entire power station. If power station consists of n generating units where each unit is represented by m-states Markov model, then Markov model for the entire power station will have m^n states. In order to find short-term indices for entire power station, this model should be built and analyzed. As one can see it will require

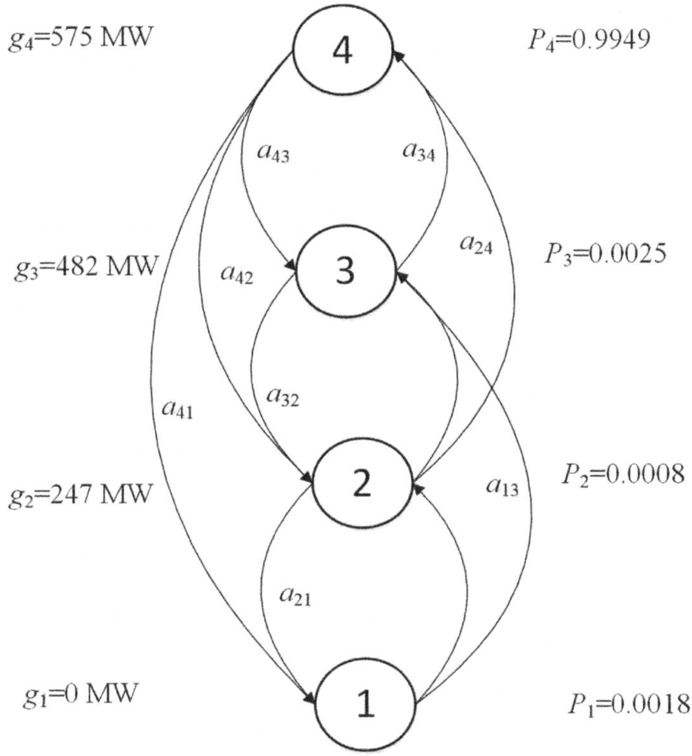

Fig. 4.1 Four-state Markov model for coal fired generating unit

enormous efforts even for relatively small m and n. In order to avoid this obstacle, we shall use L_Z-transform and illustrate corresponding benefits of its application.

4.1.2 Short-Term Reliability Analysis for Power Station with Several Coal Fired Units

In Fig. 4.2, one can see a power station, consisting of n generating units connected to common switchgear. Each generating unit i, $i = 1, 2, \ldots, n$ is describing by discrete-state continuous-time Markov process $G_i(t)$. The number of states for unit i is m_i and g_{ij} is a generating capacity of unit i in state j, $j = 1, 2, \ldots, m_i$. We designate transition rates (intensities) for transition from state c to state d for unit i as $a_{cd}^{(i)}$.

Switchgear is describing by two-state Markov model with capacities $g_{s1} = 0$ (complete failure) and $g_{s2} \geq \sum_{i=1}^{n} g_{im_i}$, and transition rates $a_{12}^{(s)}, a_{21}^{(s)}$.

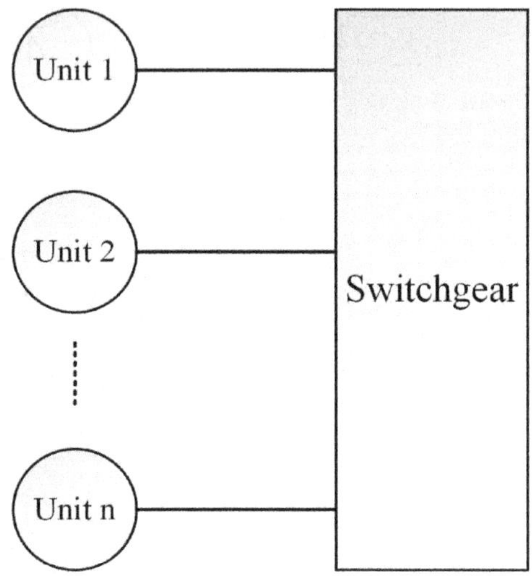

Fig. 4.2 Power system consisting of n generating units connected to switchgear

For generating unit number i one has the following system of differential equations in order to find probabilities $p_{ij}(t)(i = 1, \ldots, n; j = 1, \ldots, m_j)$, that the unit i will be in state j at instant $t \geq 0$

$$
\begin{aligned}
\frac{dp_{i1}(t)}{dt} &= a_{11}^{(i)}(t)p_{i1}(t) + a_{12}^{(i)}(t)p_{i2}(t) + \cdots + a_{im_i}^{(i)}(t)p_{im_i}(t) \\
\frac{dp_{i2}(t)}{dt} &= a_{21}^{(i)}(t)p_{i1}(t) + a_{22}^{(i)}(t)p_{i2}(t) + \cdots + a_{2m_i}^{(i)}(t)p_{im_i}(t) \\
&\ldots \\
\frac{dp_{im_i}(t)}{dt} &= a_{m_i1}^{(i)}(t)p_{i1}(t) + a_{m_i2}^{(i)}(t)p_{i2}(t) + \cdots + a_{m_im_i}^{(i)}(t)p_{im_i}(t)
\end{aligned}
\tag{4.2}
$$

The system (4.2) should be solved under specified initial conditions $\mathbf{p_{j0}}$.

After solving system (4.2) individual L_Z-transforms can be found for each Markov process $G_i(t)$, associated with generating capacity of unit i

$$L_Z\{G_i(t)\} = \sum_{j=1}^{m_i} p_{ij}(t) z^{g_{ij}}, \; i = 1, \ldots, n \tag{4.3}$$

Analogously, L_Z-transform for Markov process $G_S(t)$ associated with switchgear can be obtained

$$L_Z\{G_S(t)\} = \sum_{j=1}^{2} p_{s_j}(t) z^{g_{s_j}} = p_{s_1} z^{g_{s_1}} + p_{s_2} z^{g_{s_2}} \tag{4.4}$$

At the next step, based on these individual L_Z-transforms, L_Z-transform $L_Z\{G_Y(t)\}$ of resulting Markov process $G_Y(t)$ should be calculated by applying Ushakov's Universal Generating Operator (UGO) to all individual L_Z-transforms $L_Z\{G_j(t)\}$ over all time points $t \geq 0$

$$L_Z\{G_Y(t)\} = \Omega_f(L_Z\{G_1(t)\}, \ldots, L_Z\{G_n(t)\}, L_Z\{G_S(t)\}) \tag{4.5}$$

where f is the power system structure function.

For power system, which is depicted in Fig. 4.2, the structure function is as follows:

$$f = \min\{G_1(t) + \cdots + G_n(t), G_S(t)\} \tag{4.6}$$

because all units are connected in parallel and switchgear connected in series with all these units.

Therefore, in according to [11] we obtain

$$L_Z\{G_Y(t)\} = \Omega_{f_{ser}}\left(\Omega_{f_{par}}(L_Z\{G_1(t)\}, \ldots, L_Z\{G_n(t)\}), L_Z\{G_S(t)\}\right) \tag{4.7}$$

where $\Omega_{f_{ser}}$ and $\Omega_{f_{par}}$ are Ushakov's generating operators for elements connected in series and in parallel respectively. Calculation $\Omega_{f_{par}}$ and $\Omega_{f_{ser}}$ from (4.7) are as follows:

$$\begin{aligned}\Omega_{f_{par}}(L_Z\{G_1(t)\}, \ldots, L_Z\{G_n(t)\}) &= L_Z\{G_1(t)\} \cdot \cdots \cdot L_Z\{G_n(t)\} = \\ &= \prod_{i=1}^{n} \sum_{j=1}^{m_i} p_{ij}(t) z^{g_{ij}} = \sum_{i=1}^{r} q_i(t) z^{x_i};\end{aligned} \tag{4.8}$$

$$\Omega_{f_{ser}}\left(\sum_{i=1}^{r} q_i(t) z^{x_i}, \sum_{j=1}^{2} p_{s_j}(t) z^{g_{s_j}}\right) = \sum_{i=1}^{r} \sum_{j=1}^{2} q_i(t) p_{s_j}(t) z^{\min\{x_i, g_{s_j}\}} \tag{4.9}$$

Therefore, L_Z-transform for output stochastic process $Y(t)$ will be the following:

$$L_Z\{G_Y(t)\} = \sum_{i=1}^{r} \sum_{j=1}^{2} q_i(t) p_{s_j}(t) z^{\min\left\{x_i, g_{s_j}\right\}} = \sum_{k=1}^{K} p_k(t) z^{y_k}. \tag{4.10}$$

Now, when L_Z-transform (4.10) $L_Z\{G_Y(t)\} = \sum_{k=1}^{K} p_k(t) z^{y_k}$ of the entire MSS's output stochastic process $Y(t) \in \{y_1, \ldots, y_K\}$ is known, important availability and performability measures can be easily found.

The power station availability for demand level w is treated as system ability to provide power supply to consumers with summarized load w. It means that power station should be in states with generating capacity more or equal w.

So, the system availability for the constant demand level w at instant $t \geq 0$

$$A_w(t) = \sum_{y_i \geq w} p_i(t). \tag{4.11}$$

In other words, in order to find MSS's instantaneous availability one should summarize all probabilities in L_Z-transform from terms where powers of z are greater or equal to demand w.

The value $1 - A_w(t)$ in the literature devoted to power system is named as a loss of load probability ($LOLP_w$) for a given demand level w

$$LOLP_w(t) = 1 - A_w(t). \tag{4.12}$$

The system's expected generating capacity deficiency is expressed as follows:

$$ECD_w(t) = \sum_{k=1}^{K} p_k(t)(w - g_k)1(w - g_k), \tag{4.13}$$

where

$$1(w - g_k) = \begin{cases} 1, & \text{if } (w - g_k) > 0 \\ 0, & \text{if } (w - g_k) \leq 0 \end{cases}. \tag{4.14}$$

Based on the calculated functions $ECD_w(t)$, expected energy not supplied to consumers during time t, $EENS_w(t)$, can be computed:

$$EENS_w(t) = \int_0^t ECD(u)du. \tag{4.15}$$

Measures (4.13)–(4.15) characterize power system performability.

All measures strongly depend on the initial conditions, under which should be solved the system of differential equations for each generating unit. In other words, reliability measures for power system depend on units' initial states.

4.1.3 Short-Term Analysis for Power Plant with Three Coal Fired Generating Units

As an example of application of the method presented in Sect. 4.1.2, we shall solve here the following real-world problem. Consider a power plant (station) consisting of 3 coal fired generating units U1, U2 and U3, connected to common switchgear. U1 has nominal generating capacity 580 MW, U2 and U3 are the same and each

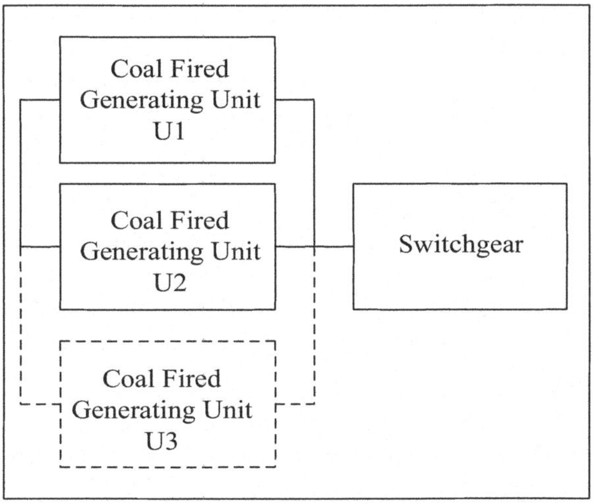

Fig. 4.3 Reliability block diagram for the power system

has nominal generating capacity 400 MW. Reliability block diagram for the power station is presented in Fig. 4.3.

U1 is presented by 4 states Markov model with corresponding capacity levels $g_{11} = 0$, $g_{12} = 300$ MW, $g_{13} = 480$ MW, $g_{14} = 580$ MW and transition intensities matrix

$$\mathbf{A}_1 = \left|a_{ij}^{(1)}\right| = \begin{vmatrix} -0.0289 & 0.0289 & 0 & 0 \\ 0.1628 & -0.4652 & 0.2791 & 0.0233 \\ 0 & 0.0235 & -0.2019 & 0.1784 \\ 0.0001 & 0.0001 & 0.0008 & -0.0010 \end{vmatrix}. \tag{4.16}$$

Each element in matrix $\mathbf{A_1}$ is represented by such units as 1/h.

Units U2 and U3 are the same and presented by 4 states Markov models with corresponding capacity levels $g_{21} = g_{31} = 0$, $g_{22} = g_{32} = 200$ MW, $g_{23} = g_{33} = 310$ MW, $g_{24} = g_{34} = 400$ MW and corresponding transition intensities matrices

$$\mathbf{A_2} = \mathbf{A_3} = \left|a_{ij}^{(2)}\right| = \begin{vmatrix} -0.1059 & 0.1059 & 0 & 0 \\ 0.0291 & -0.1456 & 0.0971 & 0.0194 \\ 0 & 0.0173 & -0.2890 & 0.2717 \\ 0.0002 & 0.0001 & 0.0010 & -0.0013 \end{vmatrix}. \tag{4.17}$$

Each element in matrix $\mathbf{A_2}$ and $\mathbf{A_2}$ is represented by such units as 1/h.

Switchgear is presented by binary-state model with states $g_{S_1} = 0$, $g_{S_2} = 1400$ MW and transition intensities matrix

$$\mathbf{A}_S = \left|a_{ij}^{(s)}\right| = \begin{vmatrix} -0.042 & 0.042 \\ 0.00011 & -0.00011 \end{vmatrix}. \tag{4.18}$$

Each element in matrix $\mathbf{A_S}$ is represented by such units as 1/h.

In according to long-term maintenance plan at some time instant t_0 generating unit U3 should be shut down in order to perform a preventive maintenance (This fact was emphasized in Fig. 4.3 *by dashed line).*

Required demand level $w = 690$ MW and probability (risk) of loss of load should be less than 0.05 or, in other words, the power system availability should be greater than required value $A_{690}^{(r)} = 0.95$.

The question is the following. Can unit U3 be shut down, if at time instant t_0 unit U1 is in state 4 with nominal capacity $g_{14} = 580$ MW, unit U2 is in state 2 with generating capacity $g_{22} = 200$ MW and the switchgear at time instant t_0 is in state 2? Power system operator is often faced with such a problem [7].

Solution Here one can notice that when unit U3 will be shut down, remaining generating capacity of unit U1 and unit U2 will be the following: $g_{14} + g_{22} = 780\text{ MW} > w = 690$ MW. It means that formally generating capacity of the power system is enough to provide the required demand. *The question is in corresponding risk of loss of load, which should be estimated only by dynamic (short-time) reliability evaluation.* Time instant $t_0 = 0$ will be determined as initial point for our calculation.

In according to L_Z-transform method at first step states probabilities $p_{ij}(t)$, $i = 1, 2$; $j = 1, 2, 3, 4$ of states j at any time instant $t \geq 0$ for generating units U1 ($i = 1$) and U2 ($i = 2$) can be found by solving the following systems of differential equations (4.19):

$$
\begin{aligned}
\frac{dp_{i1}(t)}{dt} &= -(a_{12}^{(i)} + a_{13}^{(i)} + a_{14}^{(i)})p_{i1}(t) + a_{21}^{(i)}p_{i2}(t) + a_{31}^{(i)}p_{i3}(t) + a_{41}^{(i)}p_{i4}(t)\\
\frac{dp_{i2}(t)}{dt} &= a_{12}^{(i)}p_{i1}(t) - (a_{21}^{(i)} + a_{23}^{(i)} + a_{24}^{(i)})p_{i2}(t) + a_{32}^{(i)}p_{i3}(t) + a_{42}^{(i)}p_{i4}(t)\\
\frac{dp_{i3}(t)}{dt} &= a_{13}^{(i)}p_{i1}(t) + a_{23}^{(i)}p_{i2}(t) - (a_{31}^{(i)} + a_{32}^{(i)} + a_{34}^{(i)})p_{i3}(t) + a_{43}^{(i)}p_{i4}(t)\\
\frac{dp_{i4}(t)}{dt} &= a_{14}^{(i)}p_{i1}(t) + a_{12}^{(i)}p_{i2}(t) + a_{13}^{(i)}p_{i3}(t) - (a_{41}^{(i)} + a_{42}^{(i)} + a_{43}^{(i)})p_{i4}(t)
\end{aligned}
\tag{4.19}
$$

(Unit 3 is shutting down, therefore the system of differential equations for this unit is not considered).

For $i = 1$ the system should be solved under initial conditions $p_{11}(0) = p_{12}(0) = p_{13}(0) = 0$, $p_{14}(0) = 1$ and for $i = 2$ the system should be solved under initial conditions $p_{21}(0) = p_{23}(0) = p_{24}(0) = 0$, $p_{22}(0) = 1$, because at time instant t_0 unit U1 is in state 4 and unit U2 is in state 2. From now on, we shall write initial conditions as a corresponding vector of initial probabilities. For example, if unit U1 at time instant $t = 0$ will be in state 4, we shall write the following: U1 [0001]. If unit U2 at instant $t = 0$ will be in state 2 we shall write the following: U2 [0100].

Based on probabilities $p_{ij}(t)$ that are found by solving of corresponding system of differential equations (4.19) under the given initial conditions, L_Z-transforms for generating capacity processes $G_1(t)$ and $G_2(t)$ associated with unit U1 and unit U2 respectively can be obtained

$$L_Z\{G_1(t)\} = p_{11}(t)z^0 + p_{12}(t)z^{300} + p_{13}(t)z^{480} + p_{14}(t)z^{580}$$
$$L_Z\{G_2(t)\} = p_{21}(t)z^0 + p_{22}(t)z^{200} + p_{23}(t)z^{310} + p_{24}(t)z^{400} \quad (4.20)$$

States probabilities $p_{S_j}(t)$, $j = 1, 2$ of states j at any time instant $t \geq 0$ for switchgear can be found by solving the following systems of differential equations (4.21):

$$\frac{dp_{S_1}(t)}{dt} = -a_{12}^{(s)} p_{S_1}(t) + a_{21}^{(i)} p_{S_2}(t)$$
$$\frac{dp_{S_2}(t)}{dt} = a_{12}^{(s)} p_{S_1}(t) - a_{21}^{(s)} p_{S_2}(t) \quad (4.21)$$

The switchgear at time instant t_0 is in state 2 and, therefore, the last system of differential equations should be solved under the following initial conditions: $p_{S_1}(0) = 0$, $p_{S_2}(0) = 1$. After solving of this system L_Z-transform for stochastic process $G_S(t)$ associated with switchgear can be written

$$L_Z\{G_S(t)\} = p_{S_1}(t)z^0 + p_{S_2}(t)z^{1400}. \quad (4.22)$$

The system structure function is as follows:

$$f = \min\{G_1(t) + G_2(t), G_S(t)\},$$

and, therefore, in according to (4.7) we obtain L_Z-transform for output stochastic capacity process $G_Y(t)$ as the following

$$L_Z\{G_Y(t)\} = \Omega_{f_{ser}}\left(\Omega_{f_{par}}(L_Z\{G_1(t)\}, L_Z\{G_2(t)\}), L_Z\{G_S(t)\}\right).$$

Performing computation accordingly with expressions (4.8), (4.9) and (4.10) we obtain L_Z-transform of output process $G_Y(t)$

$$\begin{aligned}L_Z\{G_Y(t)\} = {} & p_{S_2}(t)p_{14}(t)p_{24}(t)z^{980} + p_{S_2}(t)p_{14}(t)p_{23}(t)z^{890} + p_{S_2}(t)p_{13}(t)p_{24}(t)z^{880} + \\ & + p_{S_2}(t)p_{13}(t)p_{23}(t)z^{790} + p_{S_2}(t)p_{14}(t)p_{22}(t)z^{780} + p_{S_2}(t)p_{12}(t)p_{24}(t)z^{700} + \\ & + p_{S_2}(t)p_{13}(t)p_{22}(t)z^{680} + p_{S_2}(t)p_{12}(t)p_{23}(t)z^{610} + p_{S_2}(t)p_{14}(t)p_{21}(t)z^{580} + \\ & + p_{S_2}(t)[p_{12}(t) + p_{11}(t)]p_{22}(t)z^{500} + p_{S_2}(t)p_{13}(t)p_{21}(t)z^{480} + p_{S_2}(t)p_{11}(t)p_{24}(t)z^{400} + \\ & + p_{S_2}(t)p_{11}(t)p_{23}(t)z^{310} + p_{S_2}(t)p_{12}(t)p_{21}(t)z^{300} + p_{S_2}(t)p_{11}(t)p_{22}(t)z^{200} \\ & + [p_{S_2}(t)p_{11}(t)p_{21}(t) + p_{S_1}(t)]z^0. \end{aligned} \quad (4.23)$$

In according to (4.11) we obtain the power system availability for demand level $w = 690$ MW

$$\begin{aligned}A_w(t) = {} & p_{S_2}(t)p_{14}(t)p_{24}(t) + p_{S_2}(t)p_{14}(t)p_{23}(t) + p_{S_2}(t)p_{13}(t)p_{24}(t) + \\ & + p_{S_2}(t)p_{13}(t)p_{23}(t) + p_{S_2}(t)p_{14}(t)p_{22}(t) + p_{S_2}(t)p_{12}(t)p_{24}(t). \end{aligned} \quad (4.24)$$

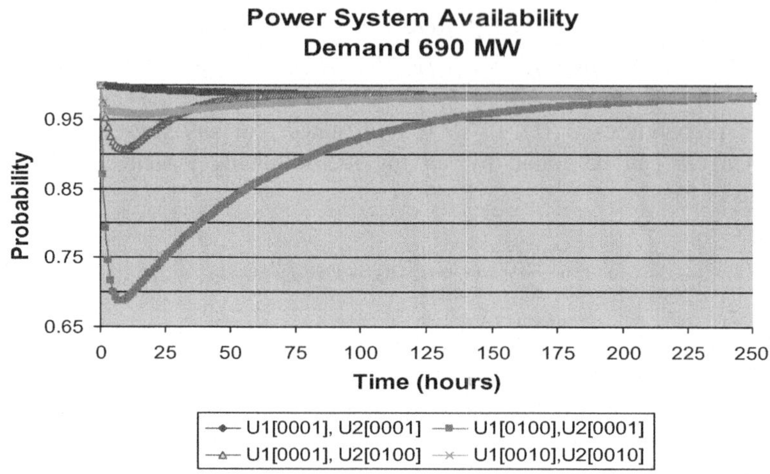

Fig. 4.4 The power system availability as function of time for different initial conditions of the units

Graphs of function $A_w(\mathrm{t})$ for different initial conditions of units U1 and U2 are presented in Fig. 4.4. There are four graphs for four different initial conditions. One can see that all other possible initial conditions cannot provide (at instant $t = t_0$) summarized generating capacity of U1 and U2 greater than required demand 690 MW, and, therefore they were not considered.

Remind that designation U1 [0001], U2 [0001] means that unit U1 and unit 2 at time instant $t = 0$ are both in the state 4; designation U1 [0100], U2 [0001] means that unit U1 at time instant $t = 0$ is in state 2 and unit U2 is in state 4; etc.

A curve that we are interested in is the curve for the given initial conditions U1 [0001], U2 [0100]. As one can see from Fig. 4.4, the required power system availability is not provided in this case. Minimum of availability function is almost 0.90 that is lower than required level $A_{690}^{(r)} = 0.95$.

***Therefore, the unit* U3 *cannot be shut down in order to perform preventive maintenance actions, if at this time instant unit* U1 *is in state 4 and unit* U2 *is in state 2. It is important conclusion for the power system operator. It means that shut down of unit* U3 *should be delayed up to improvement of initial conditions of unit* U1 *and unit* U2.**

From the Fig. 4.4, one can see that there are only two graphs that satisfy the constraint $A_{690}(t) > A_{690}^{(r)} = 0.95$. These are graphs for initial conditions U1 [0001], U2 [0001] and U1 [0010], U2 [0010]. *Therefore, the unit* U3 *can be shut down in order to perform preventive maintenance only if unit* U1 *and unit* U2 *are in these initial conditions.*

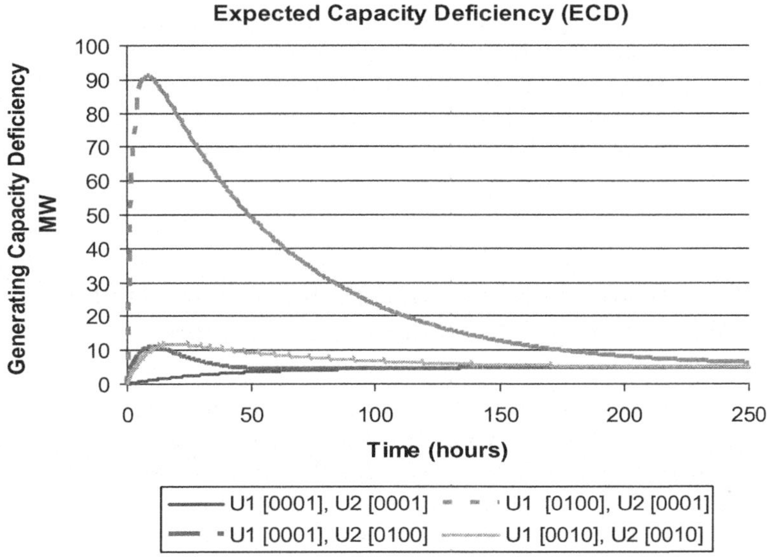

Fig. 4.5 Expected generating capacity deficiency as function of time and units' initial conditions

As one can see from Fig. 4.4 long-term power system availability

$$A_{690} = \lim_{t \to \infty} A_{690}(t) = 0.985 \tag{4.25}$$

is essentially different from calculated short-term availability, which is also strongly depends on initial conditions of the units. The transient mode is almost finished within 250 h.

In addition, by using expressions (4.13) and (4.14) one can easily compute expected capacity deficiency for all possible units' initial conditions. In Fig. 4.5 one can see the results of computation.

Maximum of function $ECD(t)$ determines the load that expected to be cut by load shedding system in order to prevent generating units' overload.

As in the previous case long-term power system ECD_{690}

$$ECD_{690} = \lim_{t \to \infty} ECD_{690}(t) = 5\ \text{MW} \tag{4.26}$$

is essentially different from calculated short-term expected capacity deficiency $ECD_{690}(t)$. For example, maximum $ECD_{690}(t)$ that was calculated for the initial conditions U1 [0100], U2 [0001] is equal 91 MW and it is much greater than long-term value 5 MW.

In this subsection we have considered L_Z-transform method application to a short-term analysis of coal fired power station. Evaluation of such important indices as power system availability, expected capacity deficiency, expected energy not supplied to consumers was considered. The method application decreases drastically a computation burden compared with straightforward Markov method. It was shown that short-term reliability indices are essentially different from long-term (steady-state) indices. Based on such computation important operational decisions can be made by power system operator. Therefore, it is worthwhile to perform such computation on-line by using corresponding advisor system.

4.2 Short-Term Availability and Performability Analysis of Power Plants with Combined Cycle Units

Combined-cycle thermal power plants are now more and more widespread. The reliability and availability of such power plants depend on the perfect operation of all its subsystems (e.g. Gas Turbine (GT), Heat Recovery Steam Generator (HRSG), Steam Turbine (ST) and cooling system). The HRSG is the link between the gas turbine and steam turbine process having the function of converting the exhaust gas energy of the gas turbine into the steam.

Combined cycle gas turbine (CCGT) generating unit may be used as a base load unit as well as a peaking unit. Peaking units normally operate for short periods of time. A basic four-state model that was suggested in [2] is often used for reliability investigation in this case. In [5] was considered the case where there are a number of CCGT units in power system and each unit is represented by a multi-state model. Because of huge dimension of the problem, only steady-state (long-term) probabilities were of interest in this work. In [4] reliability indices were introduced for combined cycle power plant by using graph theoretic approach, but representation of unit generating capacity as a stochastic process was out of the chapter scope.

4.2.1 Multi-state Markov Model for a Single Combined Cycle Generating Unit and L_Z-Transform for Its Output Generating Capacity Process

A Markov model is usually represented by a set of states in which the generating unit can reside and transitions that reflect the capability of unit to transit from one state to another in accordance with certain actions. Such state-transition diagram mimics the operating behavior of a generating unit. According to the multi-state system approach, the set of unit states should be arranged in according to the unit generating capacity.

We consider here so-called two-shaft combined cycle unit, where gas turbine generator and steam turbine generator have two different shafts. For two-shaft combined cycle unit there are three possible capacity levels. At first level, when two turbines and two generators are working, the unit capacity is the sum of nominal generating capacity of gas-turbine generator and steam turbine generator. At second level, the steam turbine is not working; therefore, the unit generating capacity is characterized only by the generating capacity of the gas turbine generator. When the gas turbine generator and the steam turbine generator are failed, then the unit generating capacity is equal to zero. We shall use the general 8-states Markov model that was suggested in [12] for two shaft combined cycle units. The corresponding state-space diagram together with all possible transitions between states and transition intensities is presented in Fig. 4.6.

The following designation will be used.

G_{gas}—nominal generating capacity of gas turbine generator,

G_{steam}—nominal generating capacity of steam turbine generator,

G_{cc}—nominal generating capacity of entire combine cycle unit; $G_{cc}= G_{gas}+ G_{steam}$,

$G(t)$—output stochastic generating capacity of entire power system at time t,

$G_i(t)$—stochastic generating capacity of unit i at time t,

$G_{i,k}$—k-th possible generating capacity level of unit i,

P_{sg}—probability of starting failure for gas turbine,

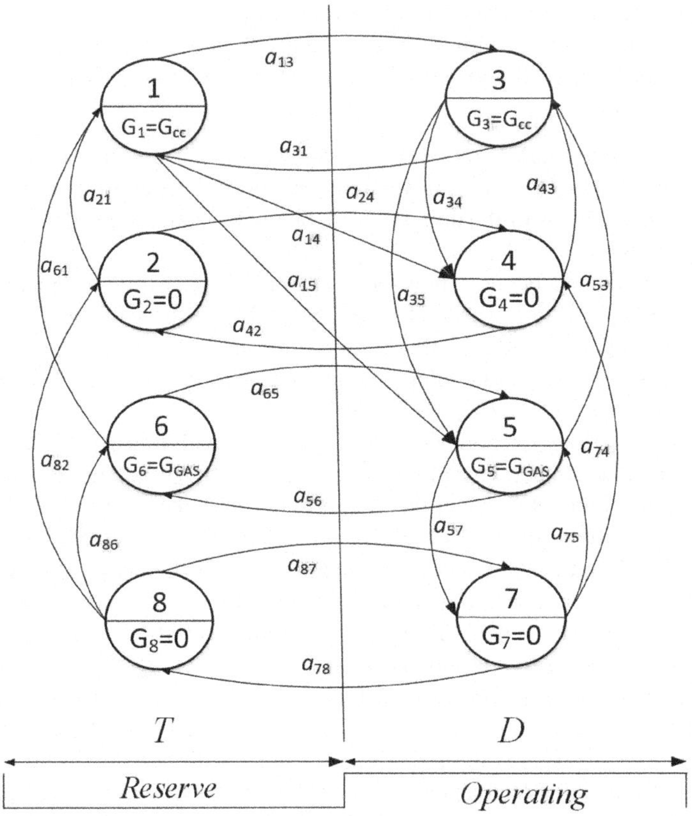

Fig. 4.6 Multi-state Markov model for a two-shaft CCGT unit

P_{ss}—probability of starting failure for steam turbine,
w—*required* power demand level for entire system,
D—average in service time per occasion of demand,
T—average reserve shutdown time between periods of need,
$MTTF_1$—average in-service time between occasions of forced outage that shutting down at gas turbine (failure of type 1),
$MTTF_2$—average in-service between occasions of forced outage that shutting down at steam turbine or steam generator (failure of type 2),
r_1—average repair time for occasions with $MTTF_1$,
r_2—average repair time for occasions with $MTTF_2$,
EENS—expected energy not supplied to consumers,
a_{ij}—intensity of transition from state i to state j,
LOLP—loss of load probability.
The model was built under the following assumptions:

- If the unit is in a reserve state (non-operating), then no failures may occur.
- The parameters Time to Failure (TTF), Time to Repair (TTR), Reserve Shutdown Time (RST) between periods of need, and in service time per occasion of demand are exponentially distributed random variables.
- Only statistical data for unit's forced outages should be considered for a short-term analysis. Planned and preventive maintenance actions should be excluded from consideration.

In the multi-state Markov model for a two-shaft CCGT unit the state 1 is reserve shutdown state where the combined cycle unit doesn't operate, because it is not needed. All components of the unit are available in this state, so available generating capacity will be $G_1 = G_{cc}$.

If a power demand occurs, then the unit transits from state 1 to state 3, if no starting failures occur nor in the GT or ST. State 3 is in-service state, where all the unit's components are available, therefore the available generating capacity in this state will be $G_3 = G_{cc}$ too. The intensity rate of transition from state 1 to state 3 will be as following $a_{13} = (1 - P_{ss})(1 - P_{sg})/T$.

If the unit is in state 3, it can transit back to state 1 with transition rate $a_{31} = 1/D$, when the unit being shut down ending its service time. If in the state 3 a forced outage occurred (failure) that shut down the gas turbine (GT), then the unit will transit to state 4 with transition rate $a_{34} = 1/\text{MTTF}_1$. In state 4 the unit's generating capacity is zero $G_4 = 0$. If in the state 3 a forced outage occurred (failure) that shut down the steam turbine (ST) or its steam generator (HRSG), then the unit will transit to state 5 with transition rate $a_{35} = 1/\text{MTTF}_2$. In state 5 the gas turbine (GT) is working, so the generating capacity in this state is $G_5 = G_{gas}$.

There are two possible transitions from the state 4. The first one is the transition from state 4 back to the state 3 with transition rate $a_{35} = 1/r_1$, when the unit has been repaired. The second one is the transition to state 2 with transition rate $a_{42} = 1/D$, when the unit shut down after a period of need before the repair will be completed. In state 2 available generating capacity of the unit is zero $G_2 = 0$. If the unit is repaired in the state 2, it will transit to state 1.

From state 5 there are 3 possible transitions. With transition rate $a_{53} = 1/r_2$ the unit will be repaired and come back to state 3. With transition rate $a_{57} = 1/MTTR_1$ the unit will transit to state 7 after a gas turbine failure. With transition rate $a_{56} = 1/D$ the unit will transit to state 6, when the unit shut down after a period of need. In state 6 generator of gas turbine is working so the available generating capacity in this state is $G_6 = G_{gas}$.

From state 7 there are three possible transitions. With transition rate $a_{75} = 1/r_1$ the unit will be repaired after a failure of type 1 and come back to state 5. With transition rate $a_{74} = 1/r_2$ the unit will be repaired after a failure of type 2 and comes to state 4. With transition rate $a_{78} = 1/D$ the unit will transit to state 8, when an operating period is finished. Available generating capacity in state 8 is zero, $G_8 = 0$, but there is no demand for the unit using in this state.

There are three possible transitions from state 8. With transition rate $a_{87} = 1/T$ the unit will come back to state 8, when reserve shutdown period ended and the unit is needed again. With transition rate $a_{82} = 1/r_2$ the unit will be repaired after a failure of type 2 and comes to state 2. Available generating capacity in state 2 is zero, $G_2 = 0$, but there is no demand for the unit in this state. With transition rate $a_{86} = 1/r_1$ the unit will be repaired after a failure of type 1 and comes to state 6. Available generating capacity in state 6, $G_6 = G_{gas}$, because gas turbine and its generator are available in this state.

There are two possible transitions from state 2. With transition rate $a_{24} = 1/T$ the unit will transit to state 4, when reserve shutdown period ended and the unit is needed again. With transition rate $a_{21} = 1/r_1$ the unit will be repaired after a failure of type 1 and transit to state 1.

There is only one possible transition from state 6. With transition rate $a_{61} = 1/r_2$ the unit will be repaired after a failure of type 2 and transit to state 1.

Therefore, the Multi-state Markov model has 8 states and 3 levels of available generating capacity: full capacity $G_{cc}= G_{gas}+ G_{steam}$, intermediate capacity G_{gas}, when only the generator of gas turbine is available, and zero capacity.

The system of differential equations (4.27) can be written for state probabilities $p_i(t)$, $i = 1, \ldots, 8$, calculation for Markov process, which state-transition diagram is depicted in Fig. 4.5.

$$\frac{dp_1(t)}{dt} = -(a_{13} + a_{14} + a_{15})p_1(t) + a_{21}p_2(t) + a_{31}p_3(t) + a_{61}p_6(t)$$
$$\frac{dp_2(t)}{dt} = -(a_{21} + a_{21})p_2(t) + a_{42}p_4(t) + a_{82}p_8(t)$$
$$\frac{dp_3(t)}{dt} = a_{13}p_1(t) - (a_{31} + a_{34} + a_{55})p_3(t) + a_{43}p_4(t) + a_{53}p_5(t)$$
$$\frac{dp_4(t)}{dt} = a_{14}p_1(t) + a_{24}p_2(t) + a_{34}p_3(t) - (a_{42} + a_{43})p_4(t) + a_{74}p_7(t)$$
$$\frac{dp_5(t)}{dt} = a_{15}p_1(t) + a_{35}p_3(t) - (a_{53} + a_{56} + a_{57})p_5(t) + a_{65}p_6(t) + a_{75}p_7(t)$$

$$\frac{dp_6(t)}{dt} = a_{56}p_5(t) - (a_{61} + a_{65})p_6(t) + a_{86}p_8(t)$$
$$\frac{dp_7(t)}{dt} = a_{57}p_5(t) - (a_{74} + a_{75} + a_{78})p_7(t) + a_{87}p_8(t)$$
$$\frac{dp_8(t)}{dt} = a_{78}p_7(t) - (a_{82} + a_{86} + a_{87})p_8(t) \quad (4.27)$$

Assume that at instant $t = 0$ the combine cycle generating unit is in state i, ($i = 1, \ldots, 8$). It means that system (4.27) should be solved under the following initial conditions

$$\mathbf{p_0} = \mathbf{p(0)} = \{p_1(0) = 0, \ldots, p_i(0) = 1, \ldots, p_8(0) = 0\}. \quad (4.28)$$

After solving the system (4.27) with the initial conditions (4.28), one obtains all states probabilities $p_i(t)$, $i = 1, \ldots, 8$, as functions of time.

When the unit is in states 1 or 3 its available capacity will be equal to G_{cc}. When the unit will be in state 5 or 6 its available capacity will equal $G_{gas\text{-}}$, and when the unit is states 2, 4, 7, or 8 its available capacity will be zero. Therefore, output stochastic generating capacity process $G(t)$ has 3 possible capacity levels: G_{cc}, G_{gas}, 0 and may be represented by the set $\mathbf{g} = \{G_1, G_2, G_3\} = \{G_{cc}, G_{gas}, 0\}$.

Therefore, the corresponding L_Z-transform of the DSCT Markov process $G(t)$, which represent available generating capacity for CCGT unit can be written as the following

$$L_Z\{G(t)\} = [p_1(t) + p_3(t)]z^{G_{cc}} + [p_5(t) + p_6(t)]z^{G_{gas}} + [p_2(t) + p_4(t) + p_7(t) + p_8(t)]z^0. \quad (4.29)$$

Therefore, at this stage Markov model for generating capacity stochastic process should be built for each CCGT generating unit in power system. Based on this model state probabilities for each generating unit are obtained as a solution of the corresponding system of differential equations (4.27) under given initial conditions (4.28) for each unit. Then individual L_Z-transform (4.29) for each generating unit can be obtained.

4.2.2 Multi-state Reliability Analysis for Power System, Consisting of Number Combine Cycle Generating Units

We consider the power system, which consists of n independent CCGT units, that is depicted in Fig. 4.7.

In this case the stochastic process $G(t)$ that represents available generating capacity of the entire power system is the sum of generating capacity processes $G_i(t)$, $i = 1, \ldots n$ of all n units in the system. The system structure function is as

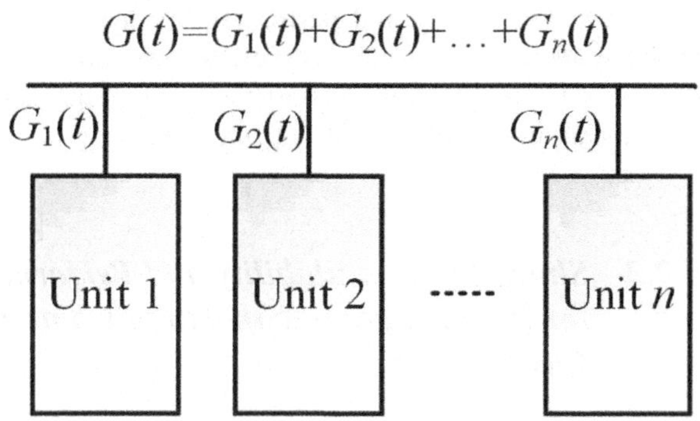

Fig. 4.7 Power system consisting of n CCGT units

follows: $f = G_1(t) + \cdots + G_n(t)$. Therefore, we have

$$G(t) = f\{G_1(t), G_2(t), \ldots, G_n(t)\} = \sum_{i=1}^{n} G_i(t) \tag{4.30}$$

In order to find the L_Z-transform of the resulting DSCT Markov process $G(t)$, which is the single-valued function $G(t) = f\{G_1(t), G_2(t), \ldots, G_n(t)\}$ of n independent DSCT Markov processes $G_j(t), j = 1, \ldots, n$, one can apply Ushakov's Universal Generating Operator (UGO) to all individual L_Z-transforms $L_Z\{G_i(t)\}$ over all time points $t \geq 0$

$$L_Z\{G(t)\} = \Omega_{f_{par}}(L_Z\{G_1(t)\}, \ldots, L_n\{G_n(t)\}) \tag{4.31}$$

In this case the resulting L_Z-transform of output generating capacity process $G(t)$ can be obtained by multiplying individual L_Z-transforms of generating capacity processes of all generating units

$$L_Z\{G(t)\} = L_Z\{G_1(t)\} \cdot \cdots \cdot L_Z\{G_n(t)\} = \sum_{i=1}^{K} p_i(t) z^{g_i} \tag{4.32}$$

For single CCGT each L_Z-transform in polynomial form $L_Z\{G_i(t)\},\ i = 1, \ldots, n$, has three terms (see expression (4.29)). Therefore, after multiplying the resulting polynomial form in expression one will have $3n$ terms. However, after combining similar terms (summarizing terms with same powers of z) the number K of terms in the resulting polynomial is usually less than $3n$.

Important availability and performability measures for the entire power system can be easily derived from the resulting L_Z-transform (4.32) by using expressions (4.11)–(4.16). We assume that the demand level w is defined as a level of load that should be met by the entire system (set of all combine cycle units).

All measures strongly depend on the initial conditions, under which the system of differential equations for each generating unit should be solved. In other words, short-term availability and performability measures for power station depend on initial states of all its units as well as on required demand level w. Below we'll demonstrate computation of short-term availability and performability indices for a power station with several CCGT units.

4.2.3 Short-Term Availability and Performability Analysis for Power Station with Three Combine-Cycle Units. Case Study

We consider a power system consisting of three independent combine cycle generating units. For such system we shall calculate availability and performability indices.

Each generating unit i, $i = 1, 2, 3$ is described by discrete-state continuous-time Markov process $G_i(t)$ with 8 states, and g_{ij} is a generating capacity of unit i in state j, $j = 1, 2, \ldots, 8$. Available generating capacity in states 1 and 3 of each CCGT unit is $G_{cc}= 370$ MW. So, we can write that $g_{ij} = 370$ MW, for $i = 1, 2, 3$ and $j = 1, 3$. Available generating capacity in states 5 and 6 for each CCGT unit is $G_{gas} = 235$ MW. So, we have $g_{ij} = 235$ MW, for $i = 1, 2, 3$ and $j = 5, 6$. Available generating capacity in states 2, 4, 7 and 8 for each CCGT unit is *ZERO*. So, we have $g_{ij} = 0$, for $i = 1, 2, 3$ and $j = 2, 4, 7, 8$.

Units' parameters are presented in Table 4.1.

The problem is to evaluate indices $A_w(t)$, $LOLP(t)$, $ECD_w(t)$ for the entire power system.

Solution At the first step system (4.16) should be written and solved for each generating unit. The solution should be performed under initial conditions that are specified for each unit.

Here we shall use the following designations for initial conditions.

Initial conditions $[i, j, k]$ means that at time instant $t = 0$ the unit 1 is in state i, unit 2 is in state j and unit 3 is in state k.

We shall measure generating capacity in any state and demand level w by using relative units. Full (total) nominal generating capacity G_{cc} of one CCGT unit, which physically is equal to 370 MW, will be equal to 1 in our computation.

Table 4.1 CCGT units' parameters

j unit number	$1/\text{MTTF}_1$ hours^{-1}	$1/r_1$ hours^{-1}	$1/\text{MTTF}_2$ hours^{-1}	$1/r_2$ hours^{-1}	1/D hours^{-1}	1/T hours^{-1}
1	0.0037	0.042	0.0014	0.042	0.0014	0.125
2	0.0045	0.039	0.0012	0.044	0.0014	0.125
3	0.0035	0.043	0.0015	0.039	0.0014	0.125

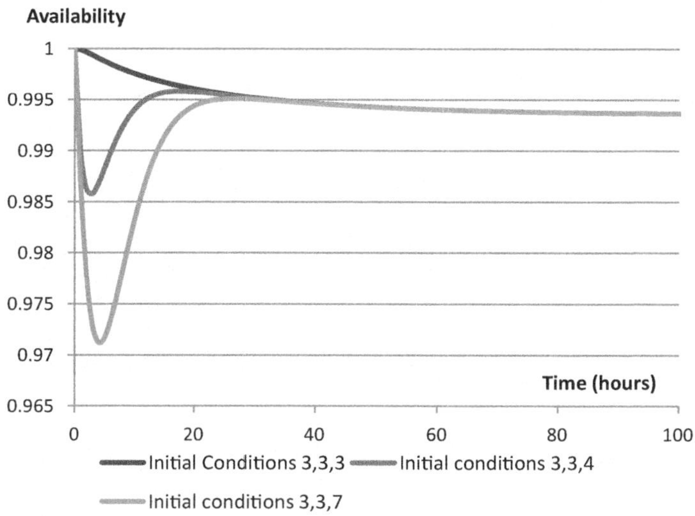

Fig. 4.8 Power system availability for demand $w = 2$

After solving all 3 systems of differential equations L_Z-transform for available generating capacity will be obtained for each unit. At the second step L_Z-transform for the entire power system can be obtained by multiplying all 3 L_Z-transforms of each unit by using expression (4.32). At the last step all short-term reliability measures for the power system that we are interested in can be derived from this L_Z-transform in according to expressions (4.22)–(4.25).

In Fig. 4.8 we present the power system availability for demand level $w = 2$ calculated for different initial conditions. From this figure one can see that long-term availability of the power system consisting of three CCGT units with total installed generating capacity $3G_{cc} = 3$ for demand level $w = 2$ is sufficiently high—$A_{w=2}(\infty) = 0.993$.

However, if, for example, two units are in perfect state 3 at instant $t = 0$ and one of the units (unit 3 in our example) will be at $t = 0$ in state 4 with zero available capacity (because of gas turbine failure), then short-term availability will drop under level 0.986 during the first 20 h. If at $t = 0$ unit 3 will be in state 7 the situation will be even worth since the availability will drop down almost to 0.97, because in state 7 gas turbine and steam turbine have failed.

Therefore, in short-term period there are substantial risks of the power system unavailability even for demand $w = 2$, *where only* 2 out of 3 *GGCT units installed in the power system have to satisfy the demand. These risks cannot be estimated from long-term reliability analysis.*

In Fig. 4.9 the power system availability for demand $w = 3$ (when all 3 units are required) is shown.

It can be seen that long-term availability $A_{w=3}(\infty) = 0.79$ which is obviously much less than long-term availability in previous case. Short-term availability also

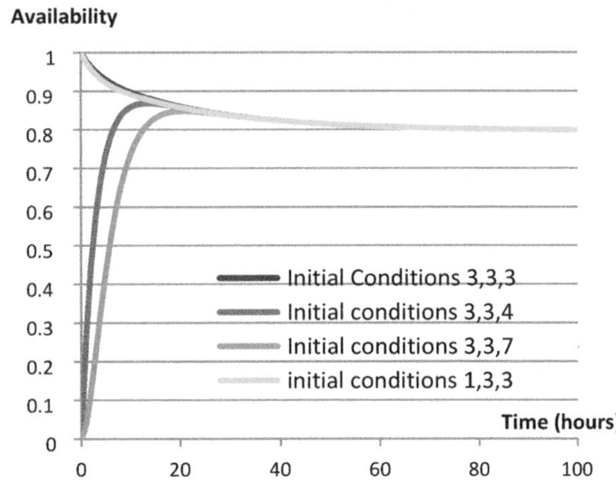

Fig. 4.9 Power system availability for demand $w = 3$

strongly depends on initial conditions. If one of the units at instant $t = 0$ is in failure state 4 or 7 (available generating capacity in both these states is zero), then the power system availability at instant $t = 0$ is zero—$A_{w=3}(0) = 0$.

Curve $A_{w=3}(t)$ for initial conditions [3, 3, 7] is under the curve $A_{w=3}(t)$ for initial conditions [3, 3, 4] because complete repair CCGT unit from state 7 is more difficult (requires repair both gas turbine and steam turbine) than from state 4 (where only gas turbine should be repaired).

The curve under initial conditions [1, 3, 3] and curve under initial conditions [3, 3, 3] are almost the same. In Fig. 4.9 these curves are positioned so close together that the difference between them cannot be distinguished visually. This illustrates the fact that for such operating mode (when CCGT unit should start approximately one time per month), probability of starting failure has no great impact on power system availability.

In Fig. 4.10 we provide the power system *LOLP* for different initial conditions.

As one can see long-term $LOLP(\infty) = 0.000013$ is much less than short-term LOLP that may be $LOLP(t = 12\,\text{h}) = 0.000092$ for initial conditions [3, 3, 7]. As in previous cases for availability computation, the worth (maximal) *LOLP* will be for initial conditions [3, 3, 7].

In Fig. 4.11 we present the power system expected capacity deficiency ECD(t) for different initial conditions of units. The required demand level is $w = 3$. It can be seen that long-term expected capacity deficiency is $A_{w=3}(\infty) = 0.16$. The curves under initial conditions [3, 3, 3] and [1, 3, 3] are so close together that the difference between them cannot be visible. As it was in the availability computation, it illustrates the fact that for such operating mode (when CCGT unit should start approximately one time per month), probability of starting failure has no a great impact on the power system *ECD*.

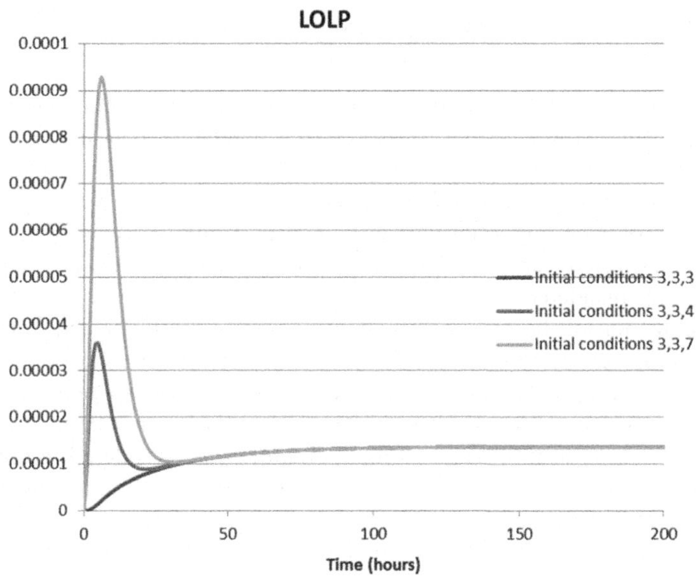

Fig. 4.10 LOLP for power system

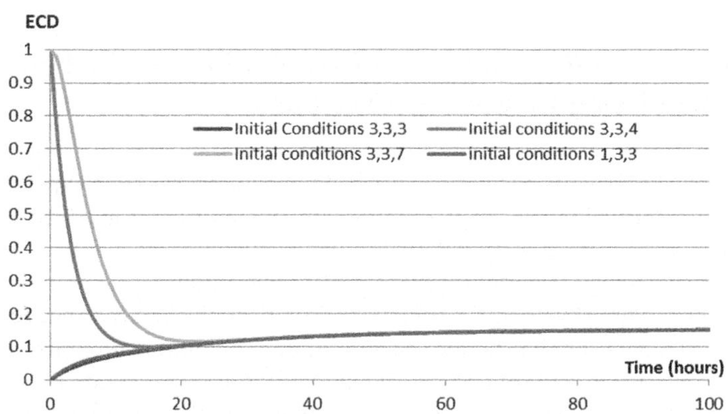

Fig. 4.11 Expected capacity deficiency (ECD) for demand $w = 3$

Expected energy not supplied to consumers can be obtained as a square under the corresponding curve $ECD_{w=3}$ (t) according to expression (4.15). Integral is considered for $t = 80$ h, because after this time the process will be in steady-state and there is no difference between curves.

Expected energy not supplied to consumers for t = 80 h for different initial conditions is as follows:

- $EENS(t = 80\text{ h}) = 9.01$ for initial conditions [3, 3, 3];
- $EENS(t = 80\text{ h}) = 9.08$ for initial conditions [1, 3, 3];

- $EENS(t = 80\text{ h}) = 12.45$ for initial conditions [3, 3, 4];
- $EENS(\text{t} = 80\text{ h}) = 14.71$ for initial conditions [3, 3, 7].

In this section is presented approach for a short-term availability and performability analysis of a power system composed by a number of non-identical combine cycle gas turbine units. Each unit is represented by a multi-state (8-state) Markov model. The approach is based on L_Z-transform.

Evaluation of such important indices as power system availability, loss of load probability, expected capacity deficiency and expected energy not supplied to consumers are considered.

The application of the proposed method decreases drastically the computation burden as compared with the straightforward Markov method.

It is shown that short-term indices are essentially different from long-term (steady-state) indices. So, operative decisions for power system cannot be made correctly when based on long-term indices. Computation method considered in this section may be used as a base for adequate control operation of power plant, which consists of several combine cycle units.

All short-term availability and performability indices of power plant with combine cycle units strongly depend on initial states of all its units as well as on required demand level w.

4.3 Short-Term Reliability and Risk Evaluation for Power System Using Inverse L_Z-Transform

In Sects. 4.1 and 4.2 of this chapter L_Z-transform technique has been applied to power system short-term evaluation. Such important indices as availability, expected energy not supplied, expected capacity deficiency were calculated. However, for effective power system dispatch it is often important to obtain power system risk function, which cannot be found by using L_Z-transform. For example, it may be needed to know how much time a system has under specified initial conditions up to its entrance in the failure state, when required power demand will not be satisfied. Therefore, it is important to know probability distribution of time up to the failure, where failure is treated as the system entrance in the set of states with unsatisfied demand. In other words, evaluation of power system risk function $Risk(t)$ is necessary. In general case reliability function $R(t)$ may be treated as $R(t) = 1 - Risk(t)$.

Based on the risk function the system operator can make appropriate operating decisions such as starting reserve generators, units shut down in order to provide maintenance and so on. For these purposes, in this section we consider an approach that is based on *inverse* L_Z–transform [8], which was mathematically defined in Chap. 3. Based on this approach, a power system risk function can be found and system operator can estimate risk corresponding to each operating decision. Generally, the suggested approach can be presented by the following steps:

1. Solving a system of ordinary differential equations for Markov model of each system element in order to obtain state probabilities as functions of time and determine L_Z-transform for each individual element.
2. Obtaining resulting L_Z-transform for the entire system output Markov process by using Ushakov's universal generating operator and corresponding techniques.
3. Uncovering underlying Markov process for the obtained resulting L_Z-transform by using inverse L_Z-transform.
4. Investigating uncovered Markov process for obtaining risk function and reliability of the entire system.

In this section we consider the application of the approach to power system risk analysis and corresponding benefits.

4.3.1 Inverse L_Z-Transform Application to Risk Function Evaluation for Power Station

In this subsection we consider a coal fired power station consisting of n generating units connected to common switchgear, which was depicted in Fig. 4.2 (see Sect. 4.1). The problem of risk function evaluation for such power system was solved in [9].

Generating capacity of each unit $i, i = 1, 2, \ldots, n$ is described by discrete-state continuous-time Markov process $G_j(t)$ as it was presented in Sect. 4.1. The number of states is m_i and g_{ij} is a generating capacity of unit i in state $j, j = 1, 2, \ldots, m_i$. We designate transition rates (intensities) for transition from state c to state d for unit i as $a_{cd}^{(i)}$.

The switchgear is described by a two-state Markov model with capacities $g_{s1} = 0$ (complete failure) and $g_{s2} = \sum_{i=1}^{n} g_{im_i}$, and transition rates $a_{12}^{(s)}, a_{21}^{(s)}$.

For generating unit number i there is the following system of differential equations in order to find probabilities $p_{ij}(t), i = 1, 2, \ldots, n; j = 1, 2, \ldots, m_i$ that the unit i will be in state j at instant $t \geq 0$

$$
\begin{aligned}
\frac{dp_{i1}(t)}{dt} &= a_{11}^{(i)}(t)p_{i1}(t) + a_{12}^{(i)}(t)p_{i2}(t) + \cdots + a_{im_i}^{(i)}(t)p_{im_i}(t) \\
\frac{dp_{i2}(t)}{dt} &= a_{21}^{(i)}(t)p_{i1}(t) + a_{22}^{(i)}(t)p_{i2}(t) + \cdots + a_{2m_i}^{(i)}(t)p_{im_i}(t) \\
&\cdots \\
\frac{dp_{im_i}(t)}{dt} &= a_{m_i1}^{(i)}(t)p_{i1}(t) + a_{m_i2}^{(i)}(t)p_{i2}(t) + \cdots + a_{m_im_i}^{(i)}(t)p_{im_i}(t)
\end{aligned}
\tag{4.33}
$$

The system (4.33) should be solved under specified initial conditions $\mathbf{p_{i0}}$ and probabilities $p_{ij}(t)$ will be found. Then individual L_Z-transforms can be written for each Markov process $G_j(t)$, associated with generating capacity of unit i,

$$L_Z\{G_i(t)\} = \sum_{j=1}^{m_i} p_{ij}(t) z^{g_{ij}}, \; i = 1, \ldots, n \tag{4.34}$$

Analogously, L_Z-transform for Markov process $G_S(t)$ associated with switchgear can be obtained

$$L_Z\{G_S(t)\} = \sum_{j=1}^{2} p_{s_j}(t) z^{g_{s_j}} = p_{s_1} z^{g_{s_1}} + p_{s_2} z^{g_{s_2}} \tag{4.35}$$

At the next step, based on these individual L_Z-transforms, L_Z-transform $L_Z\{G_Y(t)\}$ of resulting Markov process $G_Y(t)$ should be calculated by applying Ushakov's Universal Generating Operator (UGO) to all individual L_Z-transforms $L_Z\{G_j(t)\}$ over all time points $t \geq 0$

$$L_Z\{G_Y(t)\} = \Omega_f(L_Z\{G_1(t)\}, \ldots, L_Z\{G_n(t)\}, L_Z\{G_S(t)\}) \tag{4.36}$$

where f is the power system structure function.

For power system, which is depicted in Fig. 4.2, the structure function is as follows:

$$f = \min\{G_1(t) + \cdots + G_n(t), G_S(t)\} \tag{4.37}$$

because all units are connected in parallel and switchgear connected in series with all these units.

Therefore, in according to [11] we obtain

$$L_Z\{G_Y(t)\} = \Omega_{f_{ser}}\left(\Omega_{f_{par}}(L_Z\{G_1(t)\}, \ldots, L_Z\{G_n(t)\}), L_Z\{G_S(t)\}\right) \tag{4.38}$$

where $\Omega_{f_{ser}}$ and $\Omega_{f_{par}}$ are Ushakov's generating operators for elements connected in series and in parallel respectively. Calculation $\Omega_{f_{par}}$ and $\Omega_{f_{ser}}$ from (4.7) are as follows:

$$\begin{aligned} \Omega_{f_{par}}(L_Z\{G_1(t)\}, \ldots, L_Z\{G_n(t)\}) &= L_Z\{G_1(t)\} \cdot \cdots \cdot L_Z\{G_n(t)\} = \\ &= \prod_{i=1}^{n} \sum_{j=1}^{m_i} p_{ij}(t) z^{g_{ij}} = \sum_{i=1}^{r} q_i(t) z^{x_i}; \end{aligned} \tag{4.39}$$

$$\Omega_{f_{ser}}\left(\sum_{i=1}^{r} q_i(t) z^{x_i}, \sum_{j=1}^{2} p_{s_j}(t) z^{g_{s_j}}\right) = \sum_{i=1}^{r} \sum_{j=1}^{2} q_i(t) p_{s_j}(t) z^{\min\{x_i, g_{s_j}\}} \tag{4.40}$$

Finally, L_Z-transform for output stochastic process $Y(t)$ will be the following:

$$L_Z\{G_Y(t)\} = \sum_{i=1}^{r}\sum_{j=1}^{2} q_i(t)p_{s_j}(t)z^{\min\{x_i, g_{s_j}\}} = \sum_{k=1}^{K} p_k(t)z^{y_k}. \tag{4.41}$$

Now, when L_Z-transform (4.41) $L_Z\{G_Y(t)\} = \sum_{k=1}^{K} p_k(t)z^{y_k}$ of the entire MSS output stochastic process $G_Y(t) \in \{y_1, y_2, \ldots, y_K\}$ is known, then important system measures such as instantaneous availability, expected generating capacity deficiency, expected energy not supplied to consumers can be easily derived from the L_Z-transform expression.

However, risk evaluation cannot be performed based only on L_Z-transform. In order to provide risk assessment, one should uncover (reveal) the underlying stochastic process, which has L_Z-transform (4.41). For this purpose, inverse L_Z-transform (L_Z^{-1} – *transform*) should be used.

Applying inverse L_Z-transform to expression of L_Z-transform (4.41) of resulting output stochastic process one can obtain (uncover) underlying stochastic process $G_Y(t)$

$$L_Z^{-1}\{L_Z\{G_Y(t)\}\} = L_Z^{-1}\left\{\sum_{k=1}^{K} p_k(t)z^{y_k}\right\} = G_Y(t) = \{\mathbf{y}, \mathbf{A}, \mathbf{p_0}\}. \tag{4.42}$$

In multi-state interpretation a failure is defined as an event, when power system output stochastic performance (capacity) downgrades below specified (required) demand level w_{req}. Therefore, if power station output stochastic process $G_Y(t)$ is revealed (uncovered), then all states with performances lower than w_{req} should be united in one absorbing state. If $y_k < w_{req}$ and $y_{k+1} \geq w_{req}$, then all states $1, 2, \ldots, k$ should be united in one absorbing state and all transitions from this absorbing state to any other states should be constrained (forbidden). It means that in matrix $\mathbf{A}$ of the revealed Markov process all components in rows with numbers equal or lower than k should be zeroed. We designate this matrix as $\mathbf{A_0}$

$$\mathbf{A}_0 = \begin{bmatrix} 0 & 0 & \ldots & 0 \\ 0 & 0 & \ldots & 0 \\ 0 & 0 & \ldots & 0 \\ a_{k+1,1} & a_{k+1,2} & \ldots & a_{k+1,K} \\ \ldots & \ldots & \ldots & \ldots \\ a_{K1} & a_{K2} & \ldots & a_{KK} \end{bmatrix}. \tag{4.43}$$

Reliability function $.R(t).$ for the power station with required demand w_{req} can be obtained after solving the following system of differential equations under given initial conditions $\mathbf{p_{0j}}, j \geq k+1$,

$$\frac{d\mathbf{p}(t)}{dt} = \mathbf{p}(t)\mathbf{A}_0, \tag{4.44}$$

where

$\mathbf{p}(t) = [p_1(t), p_2(t), \ldots, p_K(t)]$—row-vector of state probabilities;

$\frac{d\mathbf{p}(t)}{dt} = \left[\frac{dp_1(t)}{dt}, \frac{dp_2(t)}{dt}, \ldots, \frac{dp_K(t)}{dt}\right]$—row-vector of state probabilities' first derivatives.

When row-vector of state probabilities is found, then the reliability function for the entire power station can be obtained as:

$$R_w(t) = \sum_{i=k+1}^{K} p_i(t). \tag{4.45}$$

Power station reliability function defines the probability that up to time $t > 0$ the power station does not enter the absorbing state, if at time instant $t = 0$ it was in state $j \geq k+1$, where k is defined by $w_{req} : x_k < w_{req}$ *and* $x_{k+1} \geq w_{req}$. So, a power station risk function is obtained as follows

$$Risk_w(t) = 1 - R_w(t) = 1 - \sum_{i=k+1}^{K} p_i(t) = \sum_{i=1}^{k} p_i(t). \tag{4.46}$$

Both reliability function and risk function essentially depend on initial state j. In order to emphasize this dependence, we will use the following notation:

$$Risk_w^{(j)}(t) = 1 - R_w^{(j)}(t) = 1 - \sum_{i=k+1}^{K} p_i^{(j)}(t) = \sum_{i=1}^{k} p_w^{(j)}(t), \tag{4.47}$$

where $p_i^{(j)}(t)$, $i = 1, \ldots K, j \geq k+1$, are solutions of system of differential equations (4.37) under initial conditions $\mathbf{p_{0j}}$. Note that $R_w^{(j)}(t) \equiv 0$ and $Risk_w^{(j)}(t) \equiv 1$ for $j < k+1$.

4.3.2 Risk Evaluation for Power Station with Several Coal-Fired Generating Units. Case Study

In previous sections of this chapter we considered an availability and performability evaluation for coal fired and combine-cycle power plants.

In this section, we consider a reliability and risk function computation for a coal fired power system. Here we'll follow the work [9]. We remind that risk function is treated as a probability of a power system entrance the set of unacceptable states within time period *t*.

We consider an electric power generating system, which consists of 3 coal fired generating units U1, U2 and U3 and switchgear *SW*, (see Fig. 4.2 for $n = 3$).

U1 is presented by 4 states Markov process $G_1(t) \in \{g_{11}, g_{12}, g_{13}, g_{14}\}$ with corresponding capacity levels $g_{11} = 0$, $g_{12} = 300$ MW, $g_{13} = 480$ MW, $g_{14} = 580$ MW and transition intensities matrix

$$\mathbf{A}_1 = \left|a_{ij}^{(1)}\right| = \begin{vmatrix} -0.0289 & 0.0289 & 0 & 0 \\ 0.1628 & -0.4652 & 0.2791 & 0.0233 \\ 0 & 0.0235 & -0.2019 & 0.1784 \\ 0.0001 & 0.0001 & 0.0008 & -0.0010 \end{vmatrix}. \tag{4.48}$$

Each element in matrix $\mathbf{A_1}$ is represented by such units as 1/h.

Units U2 and U3 are the same and presented by 4 states Markov processes $G_2(t) \in \{g_{21}, g_{22}, g_{23}, g_{24}\}$ and $G_3(t) \in \{g_{31}, g_{32}, g_{33}, g_{34}\}$ with corresponding capacity levels $g_{21} = g_{31} = 0$, $g_{22} = g_{32} = 200$ MW, $g_{23} = g_{33} = 310$ MW, $g_{24} = g_{34} = 400$ MW and corresponding transition intensities matrices

$$\mathbf{A_2} = \mathbf{A_3} = \left|a_{ij}^{(2)}\right| = \begin{vmatrix} -0.1059 & 0.1059 & 0 & 0 \\ 0.0291 & -0.1456 & 0.0971 & 0.0194 \\ 0 & 0.0173 & -0.2890 & 0.2717 \\ 0.0002 & 0.0001 & 0.0010 & -0.0013 \end{vmatrix}. \tag{4.49}$$

Each element in matrix $\mathbf{A_2}$ and $\mathbf{A_2}$ is represented by such units as 1/h.

Switchgear is presented by binary-state model with states $g_{S_1} = 0$, $g_{S_2} = 1400$ MW and transition intensities matrix

$$\mathbf{A}_S = \left|a_{ij}^{(s)}\right| = \begin{vmatrix} -0.042 & 0.042 \\ 0.00011 & -0.00011 \end{vmatrix}. \tag{4.50}$$

Each element in matrix $\mathbf{A_S}$ is represented by such units as 1/h.

Assume that in according to a long-term maintenance plan at time instant t_0, generating unit 3 should be shut down in order to perform a preventive maintenance. (This fact was emphasized in Fig. 4.3 by dashed line). Required demand level is the following w_{req}= 690 MW. Power system failure is treated as an event when the entire system performance downgrades below the required demand level.

At time instant t_0, unit U3 should be shut down and at that time instant unit U1 is in state 2 with nominal capacity $g_{12} = 300$ MW, unit U2 is in state 4 with generating capacity $g_{24} = 400$ MW and the switchgear at time instant t_0 is in state 2 with capacity $g_{s2} = 1400$ MW.

It means that $G_1(0) = g_{12} = 300$ MW, $G_2(0) = g_{24} = 400$ MW, $G_S(0) = g_{s2} = 1400$ MW.

The problem is to find the power system risk function in this case.

Note that formally remaining generating capacity after unit U_{-3} shut down is 300 + 400 = 700 MW, that is greater than required demand w_{req}= 690 MW. ***However, we are interested in risk associated with this situation.***

Solution Time instant t_0 will determine an initial point $t_0 = 0$ for the calculation.

We apply the algorithm described in the previous subsection.

At first step, states probabilities $p_{ij}(t),\ i = 1, 2;\ j = 1, 2, 3, 4$ of states i at any time instant $t \geq 0$ for generating units U1 ($i = 1$) and U2 ($i = 2$) should be found by solving the following systems of differential equations

$$\begin{aligned}
\frac{dp_{i1}(t)}{dt} &= -(a_{12}^{(i)} + a_{13}^{(i)} + a_{14}^{(i)})p_{i1}(t) + a_{21}^{(i)}p_{i2}(t) + a_{31}^{(i)}p_{i3}(t) + a_{41}^{(i)}p_{i4}(t)\\
\frac{dp_{i2}(t)}{dt} &= a_{12}^{(i)}p_{i1}(t) - (a_{21}^{(i)} + a_{23}^{(i)} + a_{24}^{(i)})p_{i2}(t) + a_{32}^{(i)}p_{i3}(t) + a_{42}^{(i)}p_{i4}(t)\\
\frac{dp_{i3}(t)}{dt} &= a_{13}^{(i)}p_{i1}(t) + a_{23}^{(i)}p_{i2}(t) - (a_{31}^{(i)} + a_{32}^{(i)} + a_{34}^{(i)})p_{i3}(t) + a_{43}^{(i)}p_{i4}(t)\\
\frac{dp_{i4}(t)}{dt} &= a_{14}^{(i)}p_{i1}(t) + a_{12}^{(i)}p_{i2}(t) + a_{13}^{(i)}p_{i3}(t) - (a_{41}^{(i)} + a_{42}^{(i)} + a_{43}^{(i)})p_{i4}(t)
\end{aligned} \tag{4.51}$$

For $i = 1$ the system should be solved under initial conditions $p_{11}(0) = p_{13}(0) = p_{14}(0) = 0, p_{12}(0) = 1$ and for $i = 2$ the system should be solved under initial conditions $p_{21}(0) = p_{22}(0) = p_{23}(0) = 0,\ p_{24}(0) = 1$, because at time instant t_0, the unit U1 is in state 2 and the unit U2 is in state 4. From now on in this subsection, we will write initial conditions as a corresponding vector of initial probabilities. For example, if unit U1 at time instant $t = 0$ will be in state 2, we shall write U1 [0100] keeping in mind that $G_1(0) = 300$. If the unit U2 at instant $t = 0$ will be in state 4, we shall write U2 [0001] keeping in mind that $G_2(0) = 400$.

Based on probabilities $p_{ij}(t)$ that are found by solving of corresponding system of differential equations under the given initial conditions, L_Z-transforms for generating capacity processes $G_1(t)$ and $G_2(t)$ associated with unit U1 and unit U2 respectively can be obtained

$$\begin{aligned}L_Z\{G_1(t)\} &= p_{11}(t)z^0 + p_{12}(t)z^{300} + p_{13}(t)z^{480} + p_{14}(t)z^{580}\\ L_Z\{G_2(t)\} &= p_{21}(t)z^0 + p_{22}(t)z^{200} + p_{23}(t)z^{310} + p_{24}(t)z^{400}\end{aligned} \tag{4.52}$$

States probabilities $p_{sj}(t), j = 1, 2$ of states j at any time instant $t \geq 0$ for switchgear can be found by solving the following systems of differential equations

$$\begin{aligned}\frac{dp_{s1}(t)}{dt} &= -a_{12}^{(s)}p_{s1}(t) + a_{21}^{(i)}p_{s2}(t)\\ \frac{dp_{s2}(t)}{dt} &= a_{12}^{(s)}p_{s1}(t) - a_{21}^{(s)}p_{i2}(t)\end{aligned} \tag{4.53}$$

The switchgear at time instant t_0 is in state 2. It means that $G_s(0) = 1400$. Therefore, the corresponding system of differential equations should be solved under the following initial conditions:$p_{s1}(0) = 0, p_{s2}(0) = 1$. After solving of this system L_Z-transform for capacity stochastic process $G_s(t)$ associated with switchgear can be written

$$L_Z\{G_s(t)\} = p_{s1}(t)z^0 + p_{s2}(t)z^{1400} \tag{4.53}$$

The system structure function is the following

$$f = \min\{G_1(t) + G_2(t), G_s(t)\} \tag{4.54}$$

Therefore, L_Z-transform for resulting output stochastic capacity process $G_Y(t)$ of entire MSS can be obtained

$$L_Z\{G_Y(t)\} = \Omega_{fser}\big(\Omega_{fpar}(L_Z\{G_1(t)\}, L_Z\{G_2(t)\}, L_Z\{G_S(t)\})\big) \tag{4.55}$$

Markov process $G_Y(t)$ before like term collection has 32 states.

After like term collection we obtain L_Z-transform for the entire MSS output process $G_{Y1}(t)$ that is non-Markov process (this fact was described in Chapt. 3) and has 16 states:

$$\begin{aligned}L_Z\{G_{Y1}(t)\} = {} & p_{S_2}(t)p_{14}(t)p_{24}(t)z^{980} + p_{S_2}(t)p_{14}(t)p_{23}(t)z^{890} + p_{S_2}(t)p_{13}(t)p_{24}(t)z^{880} +\\ & + p_{S_2}(t)p_{13}(t)p_{23}(t)z^{790} + p_{S_2}(t)p_{14}(t)p_{22}(t)z^{780} + p_{S_2}(t)p_{12}(t)p_{24}(t)z^{700} +\\ & + p_{S_2}(t)p_{13}(t)p_{22}(t)z^{680} + p_{S_2}(t)p_{12}(t)p_{23}(t)z^{610} + p_{S_2}(t)p_{14}(t)p_{21}(t)z^{580} +\\ & + p_{S_2}(t)\left[p_{12}(t) + p_{11}(t)\right]p_{22}(t)z^{500} + p_{S_2}(t)p_{13}(t)p_{21}(t)z^{480} + p_{S_2}(t)p_{11}(t)p_{24}(t)z^{400} +\\ & + p_{S_2}(t)p_{11}(t)p_{23}(t)z^{310} + p_{S_2}(t)p_{12}(t)p_{21}(t)z^{300} + p_{S_2}(t)p_{11}(t)p_{22}(t)z^{200}\end{aligned}$$

$$+\left[p_{S_2}(t)p_{11}(t)p_{21}(t)+p_{S_1}(t)\right]z^0. \tag{4.56}$$

It means that resulting output stochastic process $G_{Y1}(t)$ has 16 states with corresponding performance levels:

$$\begin{aligned} G_{Y1}(t) \in \mathbf{g} &= \{g_1, \ldots, g_{16}\} \\ &= \{0, 200, 300, 310, 400, 480, 500, 580, 610, 680, 700, 780, 790, 880, 890, 980\} \end{aligned} \tag{4.57}$$

State probabilities are determined by coefficients of corresponding z-powers.

As it was stated in Chap. 3 we should find an approximating Markov process $G_{1M}(t)$, *which will be equivalent to the process* $G_{Y1}(t)$ *in sense of equality all probabilities of staying in states with same performances over all time points* $t \geq 0$. In other words, we should find Markov process $G_{1M}(t)$ that will have the same L_Z-transform $L_Z\{G_{Y1}(t)\}$ with non-Markov process $G_{Y1}(t)$: $L_Z\{G_{1M}(t)\} = L_Z\{G_{Y1}(t)\}$.

By using expression for the system structure function, one can find the initial state $G_{1M}(0)$ for the resulting approximated Markov process $G_{1M}(t)$:

$$G_{1M}(0) = \min(G_1(0) + G_2(0), G_S(0)) = \min(300 + 400, 1700) = 700 \tag{4.58}$$

It means that resulting Markov process at time instant $t = 0$ is in state 11 (with performance 700 MW) and initial conditions are the following

$$\mathbf{p_{0,11}} = [0, 0, 0, 0, 0, 0, 0, 0, 0, 0, 1, 0, 0, 0, 0, 0] \tag{4.59}$$

The matrix $\mathbf{A_g}$ of transition intensities for underlying Markov process $G_{1M}(t)$ is obtained by using the Genetic Algorithm (GA). The population size of solutions in the proposed GA is 100. A binary string is used to encode a solution representing the transition intensity. The next feasible generations are produced from the one-point crossover procedure. The offspring will also mutate with probability 0.005 for avoiding premature convergence to a local optimum. The convergence criterion in the proposed GA is set as satisfying a pre-determined number of genetic cycles (500 cycles) and a number of genetic cycles without improving the solution performance (50 cycles). The GA converges to optimal solutions by performing about 420 cycles.

In (4.60) one can see all components of matrix $\mathbf{A_g}$ for the revealed approximating Markov process $G_{1M}(t)$.

$$
\mathbf{A_g} = \begin{bmatrix}
-0.70287 & 0.05063 & 0.02814 & 0.04571 & 0.04749 & 0.04869 & 0.10538 & 0..04431 & 0.02160 & 0.01638 & 0.01877 & 0.05201 & 0.02221 & 0.05230 & 0.03062 & 0.11863 \\
0.00900 & -0.10487 & 0 & 0.06288 & 0.02273 & 0 & 0.01026 & 0 & 0 & 0 & 0 & 0 & 0 & 0 & 0 & 0 \\
0 & 0.15382 & -1.04565 & 0 & 0 & 0.68602 & 0.18338 & 0.02243 & 0 & 0 & 0 & 0 & 0 & 0 & 0 & 0 \\
0.00028 & 0.03003 & 0 & -0.21673 & 0.17736 & 0 & 0 & 0 & 0.00906 & 0 & 0 & 0 & 0 & 0 & 0 & 0 \\
0.00010 & 0.00029 & 0 & 0.00097 & -0.03442 & 0 & 0 & 0 & 0 & 0 & 0.03306 & 0 & 0 & 0 & 0 & 0 \\
0.00025 & 0 & 0.07028 & 0 & 0 & -0.67273 & 0 & 0.38069 & 0 & 0.22151 & 0 & 0 & 0 & 0 & 0 & 0 \\
0.00020 & 0.44870 & 0.02353 & 0 & 0 & 0 & -1.44795 & 0 & 0.27992 & 0.62871 & 0.04244 & 0.02445 & 0 & 0 & 0 & 0 \\
0.00041 & 0 & 0 & 0 & 0 & 0.00072 & 0 & -0.15895 & 0 & 0 & 0 & 0.15782 & 0 & 0 & 0 & 0 \\
0 & 0 & 0 & 0.42637 & 0 & 0 & 0.01191 & 0 & -0.89419 & 0 & 0.20056 & 0 & 0.24686 & 0 & 0.00849 & 0 \\
0.00033 & 0 & 0 & 0 & 0 & 0.01323 & 0.02379 & 0 & 0 & -0.33773 & 0 & 0.11389 & 0.13000 & 0.05649 & 0 & 0 \\
0.00030 & 0 & 0.00050 & 0 & 0.20351 & 0 & 0.00011 & 0 & 0.00207 & 0 & -0.62580 & 0 & 0 & 0.39595 & 0 & 0.02336 \\
0 & 0.00026 & 0 & 0 & 0 & 0 & 0 & 0.07652 & 0 & 0.00141 & 0 & - & 0 & 0 & 0.24790 & 0.00872 \\
0.00031 & 0 & 0 & 0 & 0 & 0 & 0 & 0 & 0.05204 & 0.03547 & 0 & 0 & -0.71604 & 0.09906 & 0.52916 & 0 \\
0.00025 & 0 & 0 & 0 & 0 & 0.00016 & 0 & 0 & 0 & 0 & 0.06939 & 0 & 0.00097 & -0.24014 & 0 & 0.16937 \\
0.00027 & 0 & 0 & 0.00018 & 0 & 0 & 0 & 0 & 0 & 0 & 0 & 0.03659 & 0.00174 & 0 & -0.35248 & 0.31369 \\
0.00029 & 0 & 0 & 0 & 0 & 0 & 0 & 0.00032 & 0 & 0 & 0.00027 & 0.00016 & 0 & 0.0039 & 0.00103 & -0.00246
\end{bmatrix} \tag{4.60}
$$

Now the approximating Markov process $G_{1M}(t)$ was completely revealed $G_{1M}(t) = \{\mathbf{g}, \mathbf{A_g}, \mathbf{p_{0,11}}\}$.

In order to be sure that state probabilities of revealed approximating Markov process are really closed to state probabilities of primarily stochastic process that were obtained by using L_Z-transform, one can compare corresponding availability functions. It means that two functions should be compared under the same initial conditions for validating the accuracy.

The first one is availability of the entire system for demand $w = 690$ MW, that was calculated from L_Z-transform $L_Z\{G_Y(t)\}$ of original Markov process $G_Y(t)$ with 32 states before like term collection. As it was stated above the availability obtained from $L_Z\{G_Y(t)\}$ before like term collection (where $G_Y(t)$ is Markov process with 32 states) and availability obtained from $L_Z\{G_{Y1}(t)\}$ (where $G_{Y1}(t)$ is non-Markov process after like term collection) are the same. Therefore, availability for this case is as the following

$$AV_{w=690}^{(L_Z)}(t) = p_{s2}(t)p_{14}(t)p_{24}(t) + p_{s2}(t)p_{14}(t)p_{23}(t) + p_{s2}(t)p_{13}(t)p_{24}(t) + \\ + p_{s2}(t)p_{13}(t)p_{23}(t) + p_{s2}(t)p_{14}(t)p_{22}(t) + p_{s2}(t)p_{12}(t)p_{24}(t). \quad (4.61)$$

The second one is availability that was calculated for demand $w = 690$ MW, obtained from the revealed approximating Markov process $G_{1M}(t)$ with matrix $\mathbf{A_g}$ (4.60), which has 16 states:

$$G_{1M}(t) \in \mathbf{g} = \{g_1, \ldots, g_{16}\} \\ = \{0, 200, 300, 310, 400, 480, 500, 580, 610, 680, 700, 780, 790, 880, 890, 980\}$$

$$AV_{w=690}^{(L_Z^{-1})}(t) = \sum_{i=11}^{16} p_i(t) \quad (4.62)$$

where $p_i(t), i = 1, 2, \ldots, 16$ are system state probabilities with corresponding capacity (performance) g_i obtained from the revealed approximating Markov process. Only 6 states 11, 12, 13, 14, 15, 16 have capacity more than the required demand level $w = 690$ MW.

In Fig. 4.12, one can see graphs of these two availability functions, calculated under the same initial conditions: at instant $t = 0$ unit 1 is in state 2 with capacity $g_{12} = 300$ MW, unit 2 is in state 4 with capacity $g_{24} = 400$ MW and switchgear is in state 2 with capacity $g_{s2} = 1400$ MW.

As one can see, the results (availability functions) obtained from the revealed approximating Markov process and original L_Z-transform respectively, are almost the same: maximal relative error is less than 1.2%. It indicates that the suggested method has provided the revealed approximating Markov process $G_{1M}(t) = \{\mathbf{g}, \mathbf{A_g}, \mathbf{p_{0,11}}\}$ with high accuracy.

Now based on the revealed approximating Markov process $G_{1M}(t)$, one can find reliability function $R_{w=690}(t)$ and risk function.

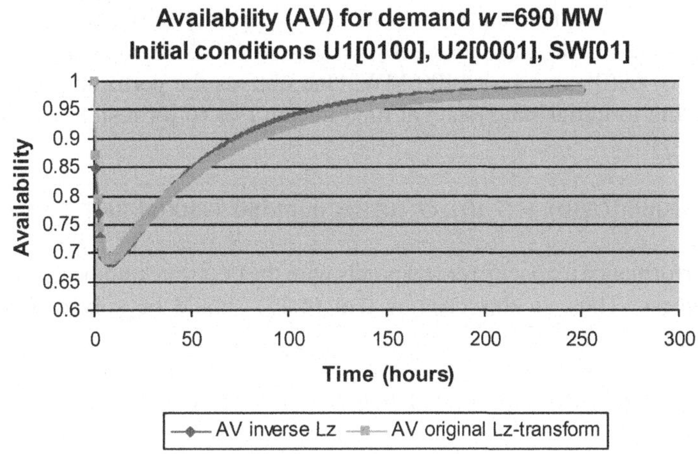

Fig. 4.12 Availability comparisons for two methods

As it was stated above for this purpose in the revealed Markov process $G_{1M}(t)$, all states $\{g_1, \ldots, g_{10}\} = \{0, 200, 300, 310, 400, 480, 500, 580, 610, 680\}$ with capacity lower than 690 MW should be united in one absorbing state.

All transitions from these absorbing states to any other states are restricted and, therefore, a new matrix $\mathbf{A_{g0}}$ will be obtained from matrix $\mathbf{A_g}$ where all components in rows with numbers equal or lower than 7 will be zeroed. Then after solving system (4.44) of differential equation the reliability function $R_{w=690}(t)$ can be obtained. (Instead of matrix $\mathbf{A_0}$, matrix $\mathbf{A_{g0}}$ is used in Eq. (4.44)).

Let us compare the reliability function for various initial conditions.

In Fig. 4.13, one can see two reliability functions with two different initial conditions. The first one was calculated for the case when all system components at instant

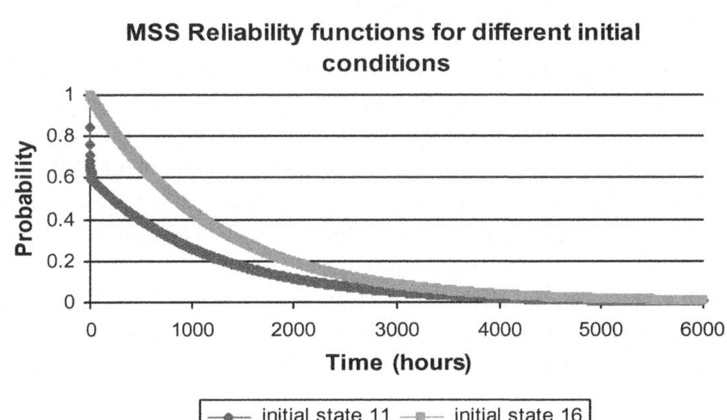

Fig. 4.13 Reliability functions $R(t)_w^{(11)}$ and $R(t)_w^{(16)}$ with two different initial conditions

$t = 0$ are in perfect states: unit U1 in state 4 with capacity $G_1(0) = g_{14} = 580$ MW, unit U2 in state 4 with capacity $G_2(0) = g_{24} = 400$ MW, switchgear in state 2 with capacity $G_s(0) = g_{s2} = 1400$ MW. One obtains the performance (capacity), corresponding to initial state (state at time instant $t = 0$) for resulting MSS output process $G_Y(t)$:

$$G_Y(0) = \min\{G_1(0) + G_2(0), G_S(0)\} = \min\{580 + 400, 1400\} = 980 \quad (4.63)$$

This performance (capacity) corresponds with the best state number 16 of resulting output process. Thus, in this case system (4.37) should be solved under initial condition

$$\mathbf{p}_{0,16}(0) = [0, 0, 0, 0, 0, 0, 0, 0, 0, 0, 0, 0, 0, 0, 0, 1] \quad (4.64)$$

The second one was calculated for the case when the system's components at instant $t = 0$ are in the following states: unit U1 is in state 2 with capacity $G_1(0) = g_{12} = 300$ MW, unit U2 is in state 4 with capacity $G_2(0) = g_{24} = 400$ MW, and switchgear is in state 2 with capacity $G_s(0) = g_{s2} = 1400$ MW. By using the system structure function, one obtains the performance (capacity), corresponding to initial state (state at time instant t = 0) for resulting MSS output process $G_Y(t)$:

$$G_Y(0) = \min\{G_1(0) + G_2(0), G_S(0)\} = \min\{300 + 400, 1400\} = 700 \quad (4.65)$$

This performance (capacity) corresponds with the state 11 of revealed output process. Therefore, in this case, system (4.44) should be solved under the following initial conditions:

$$\mathbf{p}_{0,11}(0) = [0, 0, 0, 0, 0, 0, 0, 0, 0, 0, 1, 0, 0, 0, 0, 0] \quad (4.66)$$

In Fig. 4.13, one can see two reliability functions with two different initial conditions.

As one can see, there is essential difference between two reliability functions for these various initial conditions. Function $R(t)_w^{(11)}$ is lower than $R(t)_w^{(16)}$. The difference is larger in the beginning, when t is lower than 50 h. The main reason for this is the fact that initial state 11 (capacity 700 MW) is closer to $w_{req} = 690$ MW than initial state 16 (capacity 980 MW). Therefore, when the (revealed) process $G_{1M}(t)$ begins its evolution from state 11 the probability that it falls below required capacity level w_{req} is essentially higher at the beginning.

Short-term risk functions $Risk_w^{(11)}(t)$, $Risk_w^{(16)}$ were calculated for the same two cases: when the initial state is state 11 and 16 respectively. These two functions were calculated in according to expression (4.47) for the period of 8 h. They are depicted in Fig. 4.14.

From Fig. 4.14 one can see that $Risk_w^{(11)}(t)$ is essentially greater than $Risk_w^{(16)}(t)$. (Even after 2 h $Risk_w^{(11)}(t)$ will be more than 0.25). It means that, if in the power station there is not additional reserve generating unit with quick start (for example,

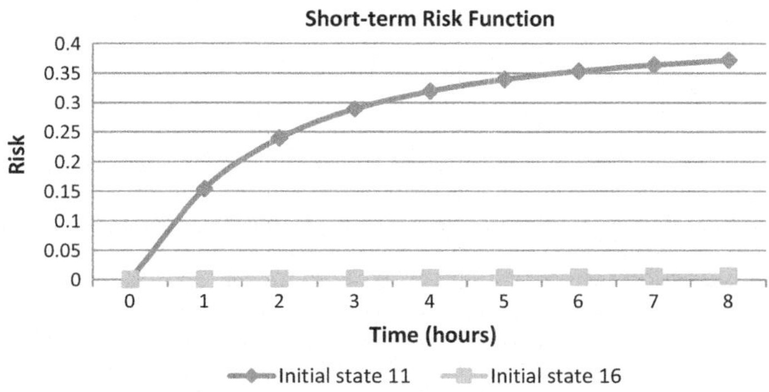

Fig. 4.14 Short-term risk functions for two different initial conditions

gas-turbine), unit U3 should not be shut down when unit U1 is in state 2 and unit U2 is in state 4, and its shut down for the preventive maintenance should be delayed.

In this section an inverse L_Z-transform (L_Z^{-1}-transform) was applied to obtaining reliability and risk function for power plant consisting of several combine-cycle units. In the proposed technique, the application of L_Z^{-1}-transform completely reveals underlying stochastic process of power station's generating capacity output, which is usually essentially simpler (has less number of states) than the process built by straightforward Markov method. Short-term risk characteristics, reliability function and mean time up to failure till now could not be found by using only L_Z-transform method. Therefore, application of the suggested technique to MSS short-term analysis expands the class of problems that can be solved. This is important in engineering practice.

4.4 Summary

In the chapter were demonstrated L_Z-transform and inverse L_Z-transform applications to short-term power system analysis.

It was shown that such important indices as Loss of Load Probability, Expected Energy Not Supplied to consumers, risk function, etc. are substantially different in short-time calculation compared with long-term (steady-steady-state) calculation.

These indices were calculated for analyzes of real-world coal fired generating units as well as for combine cycle gas turbines units. Based on these indices may be performed operative actions and decision making for units dispatch.

References

1. Billinton R, Allan R (1996) Reliability evaluation of power system. Plenum Press, New York, USA
2. Billinton R, Ge J (2004) A comparison of four-state generating unit reliability models for peaking units. IEEE Trans Power Syst 19(2):763–768
3. Billinton R, Li W (2004) Incorporating multistate unit models in composite system adequacy assessment. In: Proceedings of the international conference on probabilistic methods applied to power systems (PMAPS '04). Ames, IA, USA, 12–16 Sept 2004, pp 70–75
4. Dev N, Samsher Kachhwana SS et al (2014) Development of reliability index for combined cycle power plant using graph theoretic approach. Ain Shams Eng J 5:193–203
5. Fazekas A, Nagy E (2011) Modeling of combined heat and power generating power plant units using Markov processes with continuous time parameter and discrete state space. Electric Power Compon Syst 39:1031–1044
6. Goldner S (2006) Markov model for a typical 360 MW coal fired generation unit. Commun Depend Qual Manag 9(1):24–29
7. Lisnianski A, BenHaim H (2013) Short-term reliability evaluation for power stations by using *Lz*-transform. J Modern Power Syst Clean Energy 1(2):110–117
8. Lisnianski A, Ding Y (2014) Inverse Lz-transform for a discrete-state continuous-time Markov process and its application to multi-state system reliability analysis. In: Frenkel I, Karagrigoriou A, Lisnianski A, Kleyner A (eds) Applied reliability engineering and risk analysis: probabilistic models and statistical inference. Wiley, The Atrium, Chichester, UK, pp 43–59
9. Lisnianski A, Ding Y (2016) Using inverse L_Z-transform for obtaining compact stochastic model of complex power station for short-term risk evaluation. Reliability Eng System Saf 145:19–27
10. Lisnianski A, Elmakias D, Laredo D et al (2012) A multi-state Markov model for a short-term reliability analysis of a power generating unit. Reliab Eng Syst Safe 98(1):1–6
11. Lisnianski A, Frenkel I, Ding Y (2010) Multi-state system reliability analysis and optimization for engineers and industrial managers. Springer, London
12. Lisnianski A, Laredo D, BenHaim H (2016) Multi-state Markov model for reliability analysis of a combined cycle gas turbine power plant. In: Proceedings of the second international symposium on stochastic models in reliability engineering, life science and operations management, February 2016, Beer Sheba, Israel, pp 131–135. https://doi.org/10.1109/smrlo.2016.31
13. Reshid M, Abd Majid M (2011) A multi-state reliability model for a gas fueled cogenerated power plant. J Appl Sci 11(11):1945–1951

Chapter 5
The L_Z-Transform Application for Availability Assessment of Air Conditioning System for Chemical Laboratories

One of the principal goals of an air conditioning system is to remove or add heat and moisture from the interior of a space to improve the comfort of occupants in residential and office buildings. This process is mostly using to achieve a more comfortable interior environment; however, air conditioning is also used to cool and dehumidify rooms filled with heat-producing electronic devices, such as computer servers and power amplifiers, and to display and store delicate products, such as artwork.

Compared with refrigeration systems, which preserve products, and failure of which leads to spoilage or decreased quality of products, air conditioning systems are not as important. However, in some specific cases, such as air conditioning of chemical laboratories, equipped with exhaust hoods for hazardous materials, the technical requirements include not only temperature and humidity conditions, but also a specific rate of air replacements and constantly negative pressure in laboratory space relative to outdoor conditions. In this case, the reliability of the air conditioning system of chemical laboratories becomes a priority as it ensures personal and environmental safety of [4, 8].

In the most general sense, air conditioning can refer to any form of technology that modifies the condition of the air: heating, (de-) humidification, cooling, cleaning, ventilation, or air movement. Usually, the requirements for the reliability of air conditioning systems are not high; the reliability of such systems is estimated as 0.65–0.70 during 600 operation hours, with the lowest level of reliability having ventilation systems [9].

As opposed to air conditioning systems in residential and office buildings, air conditioning systems of chemical laboratories, and especially ventilation and air movement equipment, need to have high reliability level, to provide the personal and environment safety [12]. Increased reliability levels of such systems are achieved by redundancy of the systems' elements: exhaust systems and air handling units (AHU) with several reserved blowers, working in different reservation modes. High reliability levels and high specificity of air conditioning in chemical laboratories are

A. Lisnianski et al., *Modern Dynamic Reliability Analysis for Multi-state Systems*, Springer Series in Reliability Engineering,
https://doi.org/10.1007/978-3-030-52488-3_5

usually achieved by installing several hoods. Therefore, air conditioning systems of chemical laboratories represent redundant reliability systems with several levels of reservation [4, 11].

Air conditioning system should be regarded as a multi-state system (MSS), whose components as well as the whole system can be considered to have a finite number of states associated with various performance rates [2, 5]. The system's performance rate (output cooling capacity) can be viewed as a discrete-state continuous-time stochastic process. Such models, even in simple settings, are quite complex because they may contain several hundred states [3]. Therefore, not only the construction of such a model but also the solution of the associated system of differential equations via a straightforward Markov method is very complicated. In recent years, a special technique known as L_Z-transform has been proposed and investigated [6, 7] for discrete-state continuous-time Markov processes.

In this chapter, the L_Z-transform method is applied for the analysis of a redundant multi-state air conditioning system for chemical laboratory that must function under different cases of hood functioning. Its availability is investigated.

In Appendix, we present the MATLAB code for Availability Assessment of Air Conditioning System for Chemical Laboratory.

5.1 Description of Chemical Laboratory Air Conditioning System and Its Elements

5.1.1 Description of the System

We analyze a laboratory facilities system, used in a chemical laboratory. Although the principal goal of an air conditioning system is to keep and maintain comfortable conditions, in the case of a chemical laboratory the preference of the air conditioning system is to provide ventilation and fresh air supply, maintaining negative pressure in a laboratory in order to prevent the spread of harmful substances outside of the laboratory. The structure of the laboratory facilities system is shown in Fig. 5.1.

The system consists of a connected series (from reliability point of view): a laboratory hoods sub-system, sub-system of exhaust blowers, a sub-system of AHU blowers, and a controller. The laboratory hoods sub-system consists of five laboratory hoods. The sub-system of exhaust blowers consists of two blowers, main and reserved. The subsystem of AHU blowers consists of two blowers, main and reserved, and a proportional valve, connected in series to the blowers [1].

The primary task of the system is maintaining the negative pressure and ventilation of the laboratory using exhaust blowers. Compensation of exhausted air performs by air handling unit (AHU). An air-handling unit is an air-water heat exchanger used to regulate and circulate air as part of an air-conditioning system. Water is supplied from the central air conditioning system with a set temperature around 7 °C. The amount of water for maintaining a set point temperature in the laboratory is controlled by

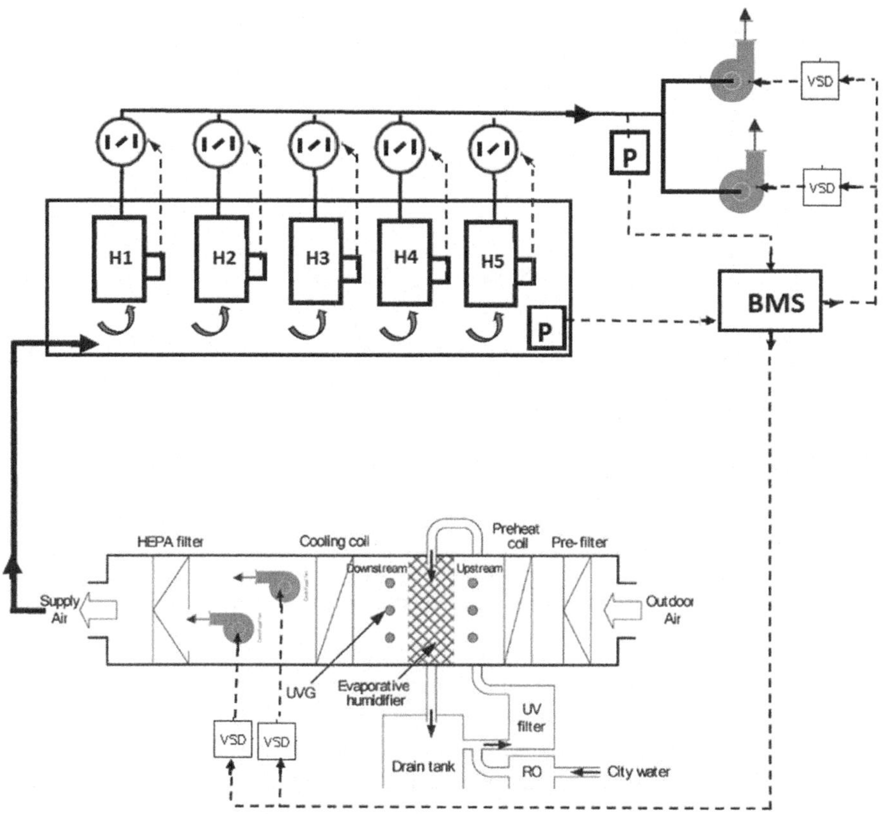

Fig. 5.1 Structure of a laboratory facilities system

a proportional control valve changing the flow area depending on the heating load. In addition, the AHU has equipment electric heating elements that allow for the air temperature to be increased if necessary. The balance between the amount of exhaust air and supplied and maintained set negative pressure is allowed by measuring the pressure in the laboratory. Based on the pressure measured, the programmable control system as a part of building management system (BMS) changes the airflow of the AHU blowers by using a variable speed device (VSD).

Flow rate changes in the exhaust system depend on the number of hoods open at the same time. The principle of operation of the chemical hood is to provide continuous airflow on the working table of the hood. To ensure safe conditions for laboratory staff, air velocity has to be about 0.6 m/s in both working states with opened door and with closed door. The given velocity is arranged by closing the air damper on the suction duct of the hood, according to the control signal from the hood controller. The signal is generated according to the speed, measured by a velocity sensor. Opening and closing of the air damper affects the air pressure in the duct, which, using VSD, change the flow rate of the exhaust blowers.

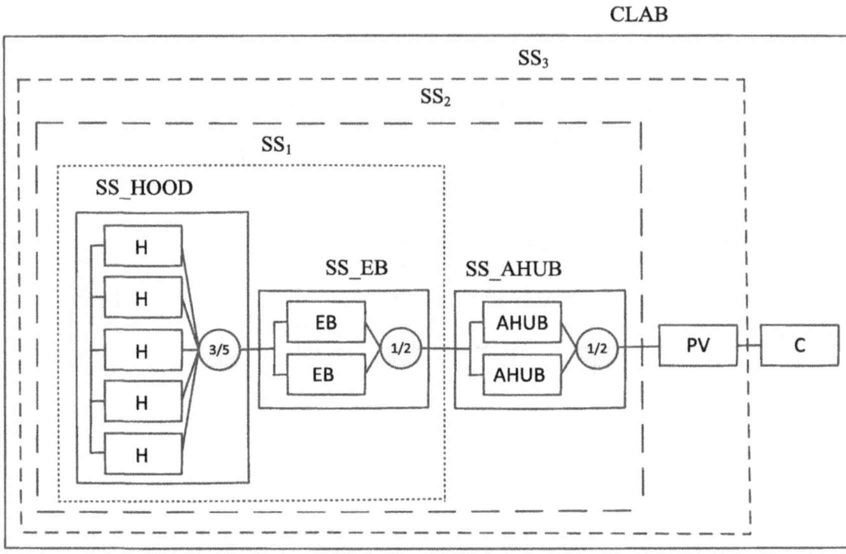

Fig. 5.2 The reliability block diagram of the air conditioning system of the chemical laboratory

The normal hood state with a closed door requires about 15–25% of maximum flow rate in comparison with the working state with an open door. In our case, the operation mode requires the possibility of three out of five available opened hoods. The maximum flow rate for each hood is 2400 m^3/h, and the maximum flow rate of each blower is 8600 m^3/h. Thus follow our operation mode one blower successfully provides stable work of three hoods. Usage of five working hood at the same time needs two simultaneously working exhaust blowers.

The air handing unit implements the temperature and humidity regime in the laboratory by replacing air removed through hoods with outside air and consists of two blowers with flow rates of 9200 m^3/h each. Similarly, the exhaust blowers have a 100% reserve for the required state of three out of five hoods, but can support the work of five hoods with two simultaneously working AHU blowers.

The proportional valve and controller support a maximum flow rate of 12,000 m^3/h for both regimes of the whole hoods sub-system.

The reliability block diagram of the air conditioning system of the chemical laboratory is presented in Fig. 5.2.

5.1.2 Description of the System's Elements

All system elements (laboratory hoods, exhaust blowers, air handing unit blowers, proportional valve, and controller) have two states: fully working and fully failed. According to the L_Z-transform method [6], in order to calculate the probabilities

Fig. 5.3 State space diagram

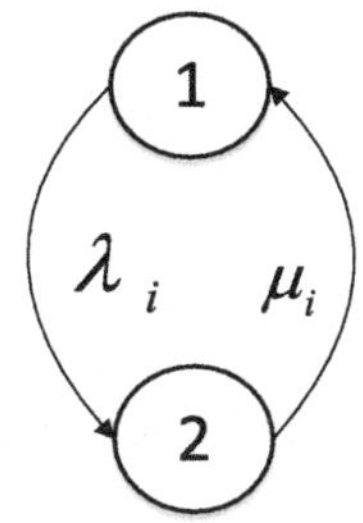

for each state, we built a state space diagram (Fig. 5.3) and the following system of differential equations [10]:

$$\begin{cases} \dfrac{dp_1^{(i)}(t)}{dt} = -\lambda_i p_1^{(i)}(t) + \mu_i p_2^{(i)}(t), \\ \dfrac{dp_2^{(i)}(t)}{dt} = \lambda_i p_1^{(i)}(t) - \mu_i p_2^{(i)}(t). \end{cases} \tag{5.1}$$

$$i = \text{H, EB, AHUB, PV, C}$$

Initial conditions are: $p_1^{(i)}(0) = 1;\ p_2^{(i)}(0) = 0.$

We used MATLAB for the numerical solution of these systems of differential equations to obtain probabilities $p_1^{(i)}(t)$, $p_2^{(i)}(t)$, (i = H, EB, AHUB, PV, C). Therefore, for the elements of such a system, the output performance stochastic processes and corresponding L_Z-transforms can be defined as follows:

For the single exhaust hood

$$\begin{cases} \mathbf{g}^{(i)} = \{g_1^H, g_2^H\} = \{2400, 0\} \\ \mathbf{p}^{(i)}(t) = \{p_1^H(t), p_2^H(t)\}. \end{cases}$$

$$L_Z\{G^H(t)\} = p_1^H(t)z^{g_1^H} + p_2^H(t)z^{g_2^H} = p_1^H(t)z^{2400} + p_2^H(t)z^0 \tag{5.2}$$

For the single exhaust blower

$$\begin{cases} \mathbf{g}^{EB} = \{g_1^{EB}, g_2^{EB}\} = \{8600, 0\} \\ \mathbf{p}^{EB}(t) = \{p_1^{EB}(t), p_2^{EB}(t)\}. \end{cases}$$

$$\begin{aligned} L_Z\{G^{EB}(t)\} &= p_1^{EB}(t)z^{g_1^{EB}} + p_2^{EB_i}(t)z^{g_2^{EB}} \\ &= p_1^{EB}(t)z^{8600} + p_2^{EB}(t)z^0 \end{aligned} \tag{5.3}$$

For the single air-handing unit blower

$$\begin{cases} \mathbf{g}^{AHUB} = \{g_1^{AHUB}, g_2^{AHUB}\} = \{9200, 0\} \\ \mathbf{p}^{AHUB}(t) = \{p_1^{AHUB}(t), p_2^{AHUB}(t)\}. \end{cases}$$

$$\begin{aligned} L_Z\{G^{AHUB}(t)\} &= p_1^{AHUB}(t)z^{g_1^{AHUB}} + p_2^{AHUB}(t)z^{g_2^{AHUB}} \\ &= p_1^{AHUB}(t)z^{9200} + p_2^{AHUB}(t)z^0 \end{aligned} \tag{5.4}$$

For proportional valve

$$\begin{cases} \mathbf{g}^{PV} = \{g_1^{PV}, g_2^{PV}\} = \{12000, 0\} \\ \mathbf{p}^{PV}(t) = \{p_1^{PV}(t), p_2^{PV}(t)\}. \end{cases}$$

$$L_Z\{G^{PV}(t)\} = p_1^{PV}(t)z^{g_1^{PV}} + p_2^{PV}(t)z^{g_2^{PV}} = p_1^{PV}(t)z^{12000} + p_2^{PV}(t)z^0. \tag{5.5}$$

For controller

$$\begin{cases} \mathbf{g}^{C} = \{g_1^{C}, g_2^{C}\} = \{12000, 0\} \\ \mathbf{p}^{C}(t) = \{p_1^{C}(t), p_2^{C}(t)\}. \end{cases}$$

$$L_Z\{G^{C}(t)\} = p_1^{C}(t)z^{g_1^{C}} + p_2^{C}(t)z^{g_2^{C}} = p_1^{C}(t)z^{12000} + p_2^{C}(t)z^0. \tag{5.6}$$

5.2 Multi-state Models for an Air Conditioning System for a Chemical Laboratory

As one can see in Fig. 5.2, the multi-state model for the air conditioning system for a chemical laboratory may be presented as connected in series laboratory hoods sub-system (SS_HOOD), sub-system of exhaust blowers (SS_EB), sub-system of air handing unit blowers(SS_AHUB), proportional valve and controller. The laboratory hoods sub-system consists of five similar laboratory hoods, connected as three-out-of-five elements, sub-systems of exhaust blowers and air handing unit blowers consist of two blowers, connected as one-out-of-two elements.

Using the recursive derivation approach [7], we will present the whole system using the L_Z-transform as follows:

$$\begin{aligned} &L_Z\left\{G^{SS\_HOOD}(t)\right\} = \Omega_{f_{3-out-of-5}}\left(L_Z\left\{g^H(t)\right\}, \left\{g^H(t)\right\}, \left\{g^H(t)\right\}, \left\{g^H(t)\right\}, \left\{g^H(t)\right\}\right), \\ &L_Z\left\{G^{SS\_EB}(t)\right\} = \Omega_{f_{1-out-of-2}}\left(L_Z\left\{g^{EB}(t)\right\}, \left\{g^{EB}(t)\right\}\right), \\ &L_Z\left\{G^{SS\_AHUB}(t)\right\} = \Omega_{f_{1-out-of-2}}\left(L_Z\left\{g^{AHUB}(t)\right\}, \left\{g^{AHUB}(t)\right\}\right), \\ &L_Z\left\{G^{SS_1}(t)\right\} = \Omega_{f_{ser}}\left(L_Z\left\{G^{HOOD}(t)\right\}, L_Z\left\{G^{EB}(t)\right\}\right) \\ &L_Z\left\{G^{SS_2}(t)\right\} = \Omega_{f_{ser}}\left(L_Z\left\{G^{SS_1}(t)\right\}, L_Z\left\{G^{AHUB}(t)\right\}\right), \end{aligned}$$

$$L_Z\left\{G^{SS_3}(t)\right\} = \Omega_{f_{ser}}\left(L_Z\left\{G^{SS_2}(t)\right\}, L_Z\left\{g^{PV}(t)\right\}\right)$$
$$L_Z\left\{G^{CLAB}(t)\right\} = \Omega_{f_{ser}}\left(L_Z\left\{G^{SS_3}(t)\right\}, L_Z\left\{g^{C}(t)\right\}\right). \quad (5.7)$$

Using the composition operators $\Omega_{f_{ser}}$, $\Omega_{f_{par}}$ and $\Omega_{f_{k-out-of-n}}$ for sub-systems, we obtain the following L_Z-transforms:

- L_Z-transforms for SS_HOOD three-out-of-five sub-system:

$$\begin{aligned}
L_Z\{G^{SS_HOOD}(t)\} &= \Omega_{f_{3-out-of-5}}\left(L_Z\{g^H(t)\}, \{g^H(t)\}, \{g^H(t)\}, \{g^H(t)\}, \{g^H(t)\}\right) \\
&= \Omega_{f_{3-out-of-5}}\left(p_1^H(t)z^{2400} + p_2^H(t)z^0\right)^5 \\
&= \Big[\{p_1^H(t)\}^5 z^{5\cdot 2400} + 5\cdot\{p_1^H(t)\}^4 p_2^H(t) z^{4\cdot 2400} \\
&\quad + 10\cdot\{p_1^H(t)\}^3\{p_2^H(t)\}^2 z^{3\cdot 2400}\Big] \\
&\quad + 10\cdot\{p_1^H(t)\}^2\{p_2^H(t)\}^3 z^{2\cdot 2400} \\
&\quad + 5\cdot p_1^H(t)\{p_2^H(t)\}^4 z^{2400} + \{p_2^H(t)\}^5 z^0 \\
&= \Big(\{p_1^H(t)\}^5 + 5\cdot\{p_1^H(t)\}^4 p_2^H(t) \\
&\quad + 10\cdot\{p_1^H(t)\}^3\{p_2^H(t)\}^2\Big) z^{\min(12000,9600,7200)} \\
&\quad + 10\cdot\{p_1^H(t)\}^2\{p_2^H(t)\}^3 z^{4800} \\
&\quad + 5\cdot p_1^H(t)\{p_2^H(t)\}^4 z^{2400} + \{p_2^H(t)\}^5 z^0 \\
&= P_1^{SS_HOOD}(t)z^{8600} + P_2^{SS_HOOD}(t)z^{4800} \\
&\quad + P_3^{SS_HOOD}(t)z^{7200} + P_4^{SS_HOOD}(t)z^0
\end{aligned} \quad (5.8)$$

where

$$\begin{aligned}
P_1^{SS_HOOD}(t) &= \{p_1^H(t)\}^5 + 5\cdot\{p_1^H(t)\}^4 p_2^H(t) + 10\cdot\{p_1^H(t)\}^3\{p_2^H(t)\}^2 \\
P_2^{SS_HOOD}(t) &= 10\cdot\{p_1^H(t)\}^2\{p_2^H(t)\}^3 \\
P_3^{SS_HOOD}(t) &= 5\cdot p_1^H(t)\{p_2^H(t)\}^4 \\
P_4^{SS_HOOD}(t) &= \{p_2^H(t)\}^5
\end{aligned}$$

- L_Z-transforms for SS_EB sub-system:

$$\begin{aligned}
L_Z\{G^{SS_EB}(t)\} &= \Omega_{f_{1-out-of-2}}\left(L_Z\{g^{EB}(t)\}, \{g^{EB}(t)\}\right) \\
&= \Omega_{f_{1-out-of-2}}\left(p_1^{EB}(t)z^{8600} + p_2^{EB}(t)z^0\right)^2
\end{aligned}$$

$$
\begin{aligned}
&= \left[\left\{p_1^{EB}(t)\right\}^2 z^{2\cdot 8600} + 2p_1^{EB}(t)\cdot p_2^{EB}(t)z^{8600}\right] + \left\{p_2^{EB}(t)\right\}^2 z^0 \\
&= \left[\left\{p_1^{EB}(t)\right\}^2 + 2p_1^{EB}(t)\cdot p_2^{EB}(t)\right] z^{\min(17200,8600)} + \left\{p_2^{EB}(t)\right\}^2 z^0 \\
&= P_1^{SS\_EB}(t)z^{8600} + P_2^{SS\_EB}(t)z^0
\end{aligned} \tag{5.9}
$$

where

$$
\begin{aligned}
&P_1^{SS\_EB}(t) = \left\{p_1^{EB}(t)\right\}^2 + 2p_1^{EB}(t)\cdot p_2^{EB}(t), \\
&P_2^{SS\_EB}(t) = \left\{p_2^{EB}(t)\right\}^2
\end{aligned}
$$

- L_Z-transforms for the SS_AHUB sub-system:

$$
\begin{aligned}
L_Z\left\{G^{SS\_AHUB}(t)\right\} &= \Omega_{f_{1-out-of-2}}\left(L_Z\left\{g^{AHUB}(t)\right\}, \left\{g^{AHUB}(t)\right\}\right) \\
&= \Omega_{f_{1-out-of-2}}\left(p_1^{AHUB}(t)z^{9200} + p_2^{AHUB}(t)z^0\right)^2 \\
&= \left[\left\{p_1^{AHUB}(t)\right\}^2 z^{2\cdot 9200} + 2p_1^{AHUB}(t)\cdot p_2^{AHUB}(t)z^{9200}\right] \\
&\quad + \left\{p_2^{AHUB}(t)\right\}^2 z^0 \\
&= \left[\left\{p_1^{AHUB}(t)\right\}^2 + 2p_1^{AHUB}(t)\cdot p_2^{AHUB}(t)\right] z^{\min(18400,9200)} \\
&\quad + \left\{p_2^{AHUB}(t)\right\}^2 z^0 \\
&= P_1^{SS\_AHUB}(t)z^{9200} + P_2^{SS\_AHUB}(t)z^0
\end{aligned} \tag{5.10}
$$

where

$$
\begin{aligned}
&P_1^{SS\_AHUB}(t) = \left\{p_1^{AHUB}(t)\right\}^2 + 2p_1^{AHUB}(t)\cdot p_2^{AHUB}(t), \\
&P_2^{SS\_EB}(t) = \left\{p_2^{AHUB}(t)\right\}^2
\end{aligned}
$$

- L_Z-transforms for the SS_1 sub-system

$$
\begin{aligned}
L_Z\left\{G^{SS_1}(t)\right\} &= \Omega_{f_{ser}}\left(L_Z\left\{G^{SS\_HOOD}(t)\right\}, L_Z\left\{G^{SS\_EB}(t)\right\}\right) \\
&= \Omega_{f_{ser}}\left(P_1^{SS\_HOOD}(t)z^{8600} + P_2^{SS\_HOOD}(t)z^{4800} + P_3^{SS\_HOOD}(t)z^{2400}\right. \\
&\quad \left. + P_4^{SS\_HOOD}(t)z^0, P_1^{SS\_EB}(t)z^{8600} + P_2^{SS\_EB}(t)z^0\right) \\
&= P_1^{SS_1}(t)^{8600} + P_2^{SS_1}(t)z^{4800} + P_3^{SS_1}(t)z^{2400} + P_4^{SS_1}(t)z^0
\end{aligned} \tag{5.11}
$$

where

$$
P_1^{SS_1}(t) = P_1^{SS\_HOOD}(t)\cdot P_1^{SS\_EB}(t)
$$

$$P_2^{SS_1}(t) = P_2^{SS\_HOOD}(t) \cdot P_1^{SS\_EB}(t)$$
$$P_3^{SS_1}(t) = P_3^{SS\_HOOD}(t) \cdot P_1^{SS\_EB}(t)$$
$$P_4^{SS_1}(t) = P_4^{SS\_HOOD}(t) \cdot P_1^{SS\_EB}(t) + P_2^{SS\_EB}(t)$$

- L_Z-transforms for the SS_2 sub-system

$$\begin{aligned} Lz\{G^{SS_2}(t)\} &= \Omega_{f_{ser}}\left(Lz\{G^{SS_1}(t)\}, Lz\{G^{SS\_AHUB}(t)\}\right) \\ &= \Omega_{f_{ser}}\Big(P_1^{SS_1}(t)z^{7200} + P_2^{SS_1}(t)z^{4800} + P_3^{SS_1}(t)z^{2400} + P_4^{SS_1}(t)z^0, \\ &\quad P_1^{SS\_AHUB}(t)z^{9200} + P_2^{SS\_AHUB}(t)z^0\Big) \\ &= P_1^{SS_2}(t)z^{7200} + P_2^{SS_2}(t)z^{4800} + P_3^{SS_2}(t)z^{2400} + P_4^{SS_2}(t)z^0 \end{aligned} \quad (5.12)$$

where

$$P_1^{SS_2}(t) = P_1^{SS_1}(t) \cdot P_1^{SS\_AHUB}(t)$$
$$P_2^{SS_2}(t) = P_2^{SS_1}(t) \cdot P_1^{SS\_AHUB}(t)$$
$$P_3^{SS_2}(t) = P_3^{SS_1}(t) \cdot P_1^{SS\_AHUB}(t)$$
$$P_4^{SS_2}(t) = P_4^{SS_1}(t) \cdot P_1^{SS\_AHUB}(t) + P_2^{SS\_AHUB}(t)$$

- L_Z-transforms for the SS_3 sub-system

$$\begin{aligned} Lz\{G^{SS_3}(t)\} &= \Omega_{f_{ser}}\left(Lz\{G^{SS_2}(t)\}, Lz\{g^{PV}(t)\}\right) \\ &= \Omega_{f_{ser}}\Big(P_1^{SS_2}(t)z^{7200} + P_2^{SS_2}(t)z^{4800} + P_3^{SS_2}(t)z^{2400} + P_4^{SS_2}(t)z^0, \\ &\quad p_1^{PV}(t)z^{12000} + p_2^{PV}(t)z^0\Big) \\ &= P_1^{SS_3}(t)z^{7200} + P_2^{SS_3}(t)z^{4800} + P_3^{SS_3}(t)z^{2400} + P_4^{SS_3}(t)z^0 \end{aligned} \quad (5.13)$$

where

$$P_1^{SS_3}(t) = P_1^{SS_2}(t) \cdot p_1^{PV}(t)$$
$$P_2^{SS_3}(t) = P_2^{SS_2}(t) \cdot p_1^{PV}(t)$$
$$P_3^{SS_3}(t) = P_3^{SS_2}(t) \cdot p_1^{PV}(t)$$
$$P_4^{SS_3}(t) = P_4^{SS_2}(t) \cdot p_1^{PV}(t) + p_2^{PV}(t)$$

- L_Z-transforms for the CLAB system

$$
\begin{aligned}
L_Z\left\{G^{CLAB}(t)\right\} &= \Omega_{f_{ser}}\left(L_Z\left\{G^{SS_3}(t)\right\}, L_Z\left\{g^{C}(t)\right\}\right) \\
&= \Omega_{f_{ser}}\Big(P_1^{SS_3}(t)z^{8600} + P_2^{SS_3}(t)z^{4800} + P_3^{SS_3}(t)z^{2400} + P_4^{SS_3}(t)z^{0}, \\
&\quad p_1^{C}(t)z^{12000} + p_2^{C}(t)z^{0}\Big) \\
&= P_1^{CLAB}(t)z^{8600} + P_2^{CLAB}(t)z^{4800} + P_3^{CLAB}(t)z^{2400} + P_4^{CLAB}(t)z^{0} \quad (5.14)
\end{aligned}
$$

where

$$
\begin{aligned}
P_1^{CLAB}(t) &= P_1^{SS_3}(t)\cdot p_1^{C}(t) \\
P_2^{CLAB}(t) &= P_2^{SS_3}(t)\cdot p_1^{C}(t) \\
P_3^{CLAB}(t) &= P_3^{SS_3}(t)\cdot p_1^{C}(t) \\
P_4^{CLAB}(t) &= P_4^{SS_3}(t)\cdot p_1^{C}(t) + p_2^{C}(t)
\end{aligned}
$$

5.3 Calculation of the Availability Indices of an Air Conditioning System for a Chemical Laboratory

The failure and repair rates for each system's elements are presented in the Table 5.1.

As one can see, single laboratory hood and proportional valve are aging equipment and their failure rates are increasing functions.

The instantaneous availability for constant demand level $w = 7200$ m^3/h may be presented as follows:

$$
A^{CLAB}(t) = \sum_{g_i^{CLAB}\geq 7200} P_i^{CLAB}(t) = P_1^{CLAB}(t). \quad (5.15)
$$

The calculated availability indices of the air conditioning system for a chemical laboratory after a year of operations are presented in Figs. 5.4, 5.5, 5.6, 5.7, 5.8 and 5.9 and Table 5.2.

Table 5.1 Failure and repair rates of elements in an air conditioning system for a chemical laboratory (year^{-1})

System's elements	Failure rates	Repair rates
Single laboratory hood	1 + 0.2t	182
Exhaust blower	6	90
Air handing unit blower	4	90
Proportional valve	2(1 + 0.2t)	121
Controller	2	182

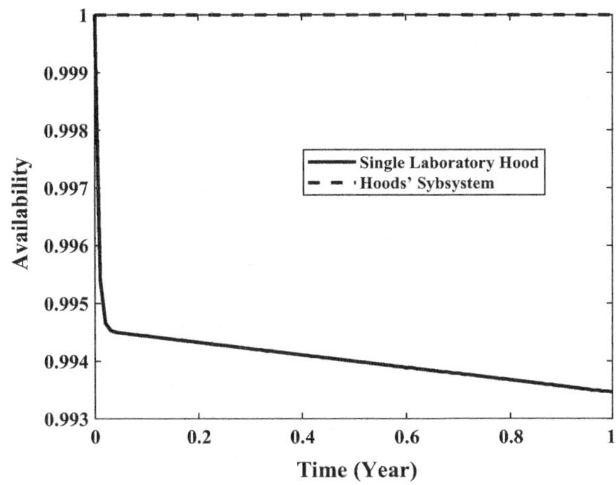

Fig. 5.4 Availability of a single laboratory hood and hoods sub-system

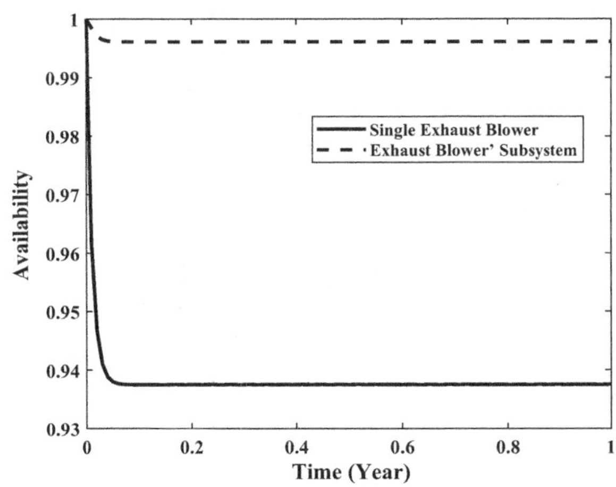

Fig. 5.5 Availability of a single exhaust blower and exhaust blower sub-system

Froma a reliability point of view the results show the correct choice of design conditions—three working hoods out of five existing, allowing high availability (0.9644) of the whole system during one year.

The highest level of availability for a hood sub-system (1.00) with the availability of a single hood 0.9935 is achieved by a redundace system as three out of five. This ensures the safety of laboratory personnel and the environment, given that in the laboratory experiments with hazardous substances are conducted.

As one can see from the resulting graphs, during a very short period, availability of elements and sub-systems reaches a steady-state regime for elements with a constant failure rate and continues to decrease for degrade elements with increasing failure rate.

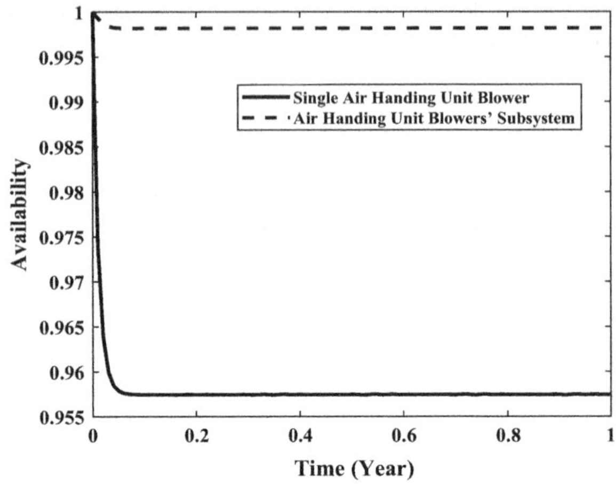

Fig. 5.6 Availability of a single air handing unit blowers and of air handing unit blower sub-system

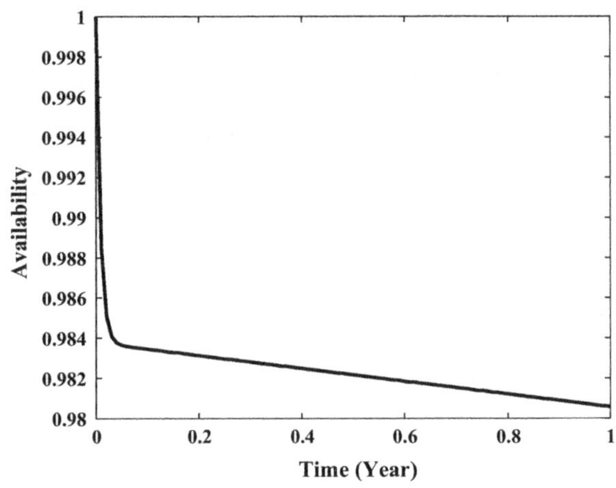

Fig. 5.7 Availability of proportional valve for chemical laboratory

5.4 Summary

In this chapter, the L_Z-transform method was used for the evaluation of important parameters of an air conditioning system for a chemical laboratory. The results of the calculations based on the L_Z-transform correlate with the statistical data on the operational availability of the air conditioning system, as discussed in the literature.

In comparison with the straightforward Markov method, which would have required the construction and solution of a model with 2048 states, the L_Z-transform approach simplifies the solution of reliability calculation for an air conditioning system for the real chemical laboratory.

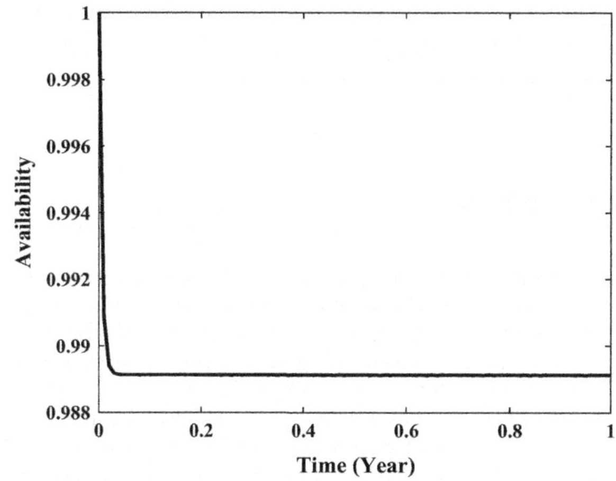

Fig. 5.8 Availability of a controller

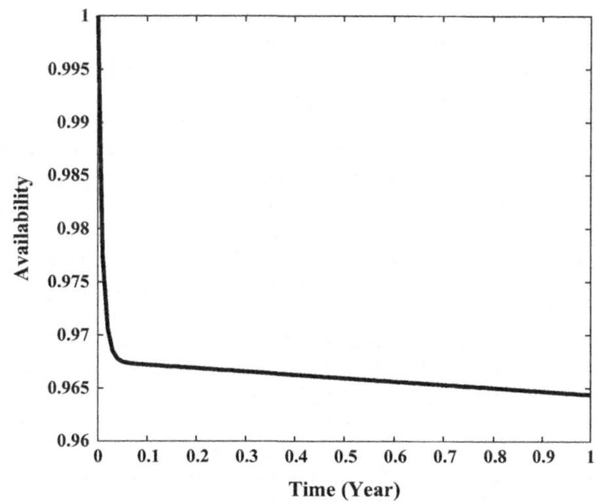

Fig. 5.9 Availability of air conditioning system for chemical laboratory

Table 5.2 Reliability indices of an air conditioning system for a chemical laboratory after a year of operations

Reliability indices	Single unit	Air conditioning system
Instantaneous availability of laboratory hood sub-system	0.9935	1.0000
Instantaneous availability of exhaust blower sub-system	0.9375	0.9961
Instantaneous availability of air handing unit blower sub-system	0.9575	0.9982
Instantaneous availability of proportional valve	0.9806	
Instantaneous availability of controller	0.9891	
Instantaneous availability of an air conditioning system for a chemical laboratory		0.9644

From a reliability point of view the results show the corect choice of design conditions—three working hood from five existing, allowed high availability (0.9644) of the whole system during one year.

References

1. ASHRAE (2018) Classification of laboratory ventilation design levels, TC 9.10, Laboratory Classification Subcommittee, Atlanta, GA 30329
2. Frenkel I, Khvatskin L, Lisnianski A (2014) Lz-transform application to availability assessment of the air conditioning system with rental equipment working under seasonal weather conditions. J Inf, Manage Control Syst 12(2):133–140
3. Frenkel I, Lisnianski A, Khvatskin L (2012) Availability assessment for aging refrigeration system by using Lz-transform. J Reliab Stat Stud 5(2):33–43
4. Jin M, Memarzadeh F, Lee KS et al (2012) Experimental study of ventilation performance in laboratories with chemical spills. Build Environ 57:327–335
5. Khvatskin L, Frenkel I (2014) On availability comparison of reservation modes for multi-state air conditioning systems using Markov approach. In: Proceedings of the 9th international conference on availability, reliability and security (ARES2014), Fribourg, Switzerland, 8–12 September 2014. IEEE CPS, pp 393–396
6. Lisnianski A (2012) Lz-transform for a Discrete-State Continuous-Time Markov process and its application to multi-state systems reliability. In: Lisnianski A, Frenkel I (eds) Recent advances in system reliability. Springer, London, pp 79–95
7. Lisnianski A, Frenkel I, Ding Y (2010) Multi-state system reliability analysis and optimization for engineers and industrial managers. Springer, London
8. Moran L, Masciangioli T (eds) (2010) Chemical laboratory safety and security. A Guide to prudent chemical management. The National Academies Press, Washington, DC
9. Myrefelt S (2004) The reliability and availability of heating, ventilation and air conditioning systems. Energy Build 36:1035–1048
10. Trivedi K (2002) Probability and statistics with reliability, queuing and computer science applications. Wiley, New York
11. Wang JJ, Fu C, Yang K et al (2013) Reliability and availability analysis of redundant BCHP (building cooling, heating and power) system. Energy 61(1):531–540
12. Zhai JZ (2005) Simulation-based ventilation system study for chemical laboratory. In: Proceedings of the 10th international conference on indoor air quality and climate, pp 3371–3376

Chapter 6
Availability and Performability of the Reserved Cold Water Supply System for a Raw Material Manufacturing for Plastics Industry

Today's production of raw materials for plastic and chemical industry requires effective cooling and precise temperature control or cooling below the ambient temperature. Removing heat from these systems efficiently is necessary to reduce 'thermal stress' which will, in turn, lead to increased accuracy, longevity and reduced energy consumption. Greater cooling efficiencies will also allow production equipment to be operated for longer periods without overheating, providing a greater return on investment for production facilities and a likely increase in availability of the equipment.

Cooling system for production of raw materials bases on cold water supply carries out the concentrated department. The raw materials are undergoing a process of mixing and processing production lines (extruders), resulting in a form of granules which gives the final product the unique features for factory's customers. In case the water cooling system does not fix the demand conditions, quality of granules will be reduced, that leads to financial lost and the extruders of the production lines may be destroying. Therefore, this type of system needs several levels of reserve.

We consider a reserved cold-water supply system for a factory, specializes in producing raw materials for the plastics industry [4]. The system consists of two close-circle parts, connected in series via differential tap. The first part is chilling system, consists of pump's sub-system and cooling water chiller, reserved by Backup sub-system. The second close-circle part consists of the pump's sub-system. The nominal cooling capacity of the entire system is $16 \cdot 10^6$ BTU/year. The cooling demand value varies from $10 \cdot 10^6$ BTU/ year to the nominal performance $16 \cdot 10^6$ BTU/year.

Due to the system's nature, a fault of a single unit has only partial effect on the entire performance: it only reduces the system's cooling capacity. Partial failure of chillers' elements, pumps or different tap leads to partial system failure (reduction of output cooling capacity) as well as to complete system failures. So, the water cooling system can be treated as multi-state system (MSS), where components and entire system in general case have an arbitrary finite number of states corresponding

A. Lisnianski et al., *Modern Dynamic Reliability Analysis for Multi-state Systems*, Springer Series in Reliability Engineering,
https://doi.org/10.1007/978-3-030-52488-3_6

to the different performance rates [2, 5–7]. The performance rate (cool capacity) of the system at any instant t is interpreted as a discrete-state continuous-time stochastic process [3, 5]. Water cooling systems often have aging components and so, one should consider in general a non-homogeneous Markov model because some transition rates (intensities) being time-dependent. Such a model is complex enough - even in relatively simple cases it has hundreds of states. Therefore, it is rather difficult to build the model and to solve the corresponding system of differential equations by using straightforward Markov method. So, in order to make a dynamic MSS reliability analysis L_Z-transform will be used.

6.1 Multi-state Model of the Reserved Cold Water Supply System

6.1.1 System's Description

The principal scheme of the cold water supply system is presented in Fig. 6.1.

Figure 6.2 presents the reliability block-diagram of the MSS cold water supply system. The system consists of 15 elements, combined into following subsystems: two identical chiller sub-systems (Ch_1 and Ch_2), three pump sub-systems (PS_1, PS_2, PS_3), and differential Tap.

The first chiller subsystem (Ch_1) consists of connected in series compressor ($Comp_1$), condenser ($Cond_1$) and evaporator (Ev_1). Similarity, the second chiller's subsystem (Ch_2) consists of connected in series compressor ($Comp_2$), condenser ($Cond_2$) and evaporator (Ev_2).

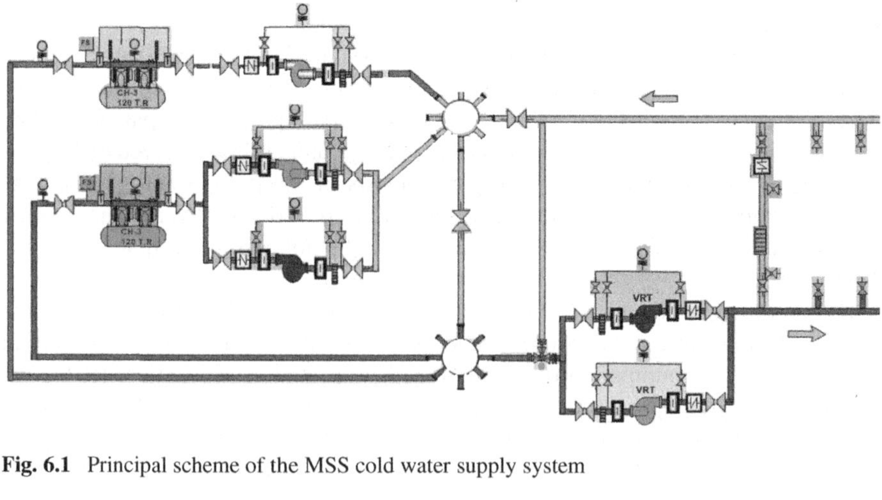

Fig. 6.1 Principal scheme of the MSS cold water supply system

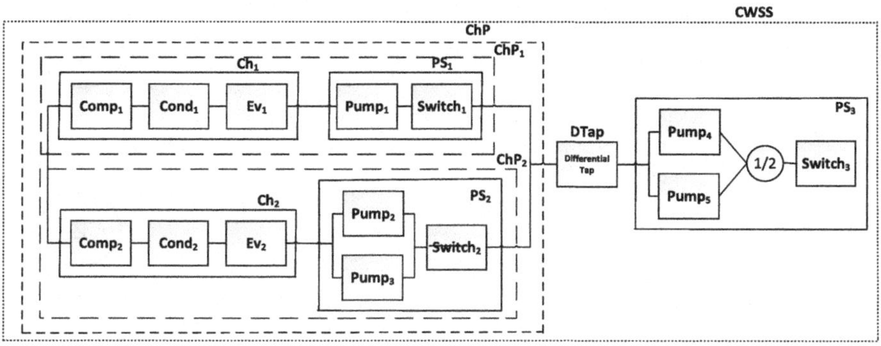

Fig. 6.2 Reliability block-diagram of the cold water supply system

The first pump's subsystem (PS_1) consists of a pump ($Pump_1$), connected in series to the switch ($Switch_1$). The second pump sub-system (PS_2) consist of two pumps ($Pump_2$ and $Pump_3$), connected in parallel and the switch. The third pump's sub-system (PS_3) consist of two pumps ($Pump_4$ and $Pump_5$) connected as one-out-of-two (cold reserve) and the switch ($Switch_3$).

For application of the recursive approach [6], we present the cold water supply system as the following sub-systems (Fig. 6.2).

- connected in series chiller sub-systems (Ch_1) and pump sub-systems (PS_1) as sub-system ChP_1;
- connected in series chiller sub-systems (Ch_2) and pump sub-systems (PS_2) as sub-system ChP_2;
- connected in parallel two sub-systems ChP_1 and ChP_2 as sub-system ChP;
- connected in series sub-system ChP, differential tap (DTap) and pump sub-systems (PS_3) as system CWSS.

6.1.2 Description of System's Elements

There are two types of system's elements. Compressors, Evaporators, Pumps and Differentia Tap are two states elements. They have working state and failed state. Condensers are four states elements (one working state, two reduce performance states and failed state). According to L_Z-transform method, described in the Chap. 3, in order to calculate the probabilities for each state, we build the state space diagrams (Figs. 6.3 and 6.4) and corresponding systems of differential equations for each element.

- For two states elements

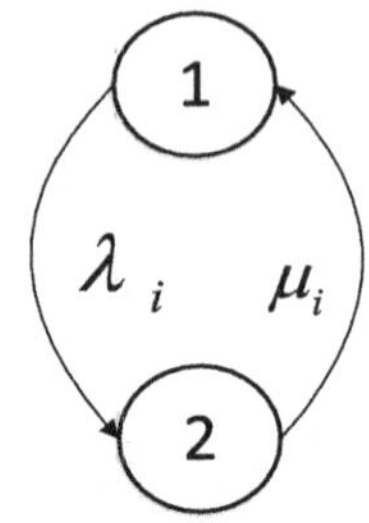

Fig. 6.3 State space diagram of two states elements

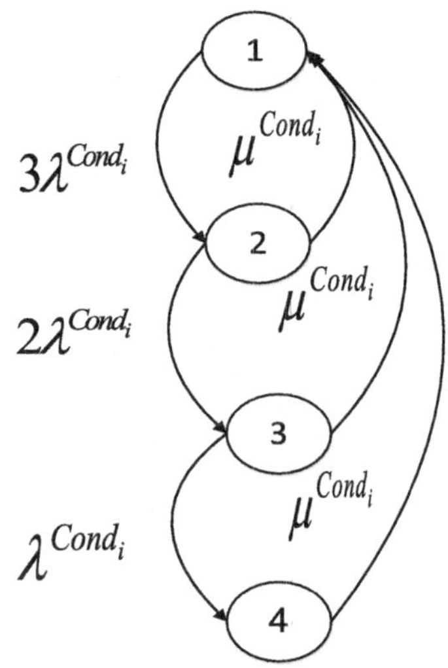

Fig. 6.4 State space diagram of four states element (condenser)

$$\begin{cases} \dfrac{dp_1^{(i)}(t)}{dt} = -\lambda_i p_1^{(i)}(t) + \mu_i p_2^{(i)}(t), \\ \dfrac{dp_2^{(i)}(t)}{dt} = \lambda_i p_1^{(i)}(t) - \mu_i p_2^{(i)}(t). \end{cases}$$

$I = \mathrm{Comp}_1, \mathrm{Comp}_2, \mathrm{Pump}_1, \ldots, \mathrm{Pump}_5, \mathrm{Ev}_1, \mathrm{Ev}_2, \mathrm{DTap}$
Initial conditions are: $p_1^{(i)}(0) = 1;\ p_2^{(i)}(0) = 0.$.

- For four states element (condenser)

$$\begin{cases} \dfrac{dp_1^{Cond_i}(t)}{dt} = -3\lambda^{Cond_i} p_1^{Cond_i}(t) + \mu^{Cond_i}\left[p_2^{Cond_i}(t) + p_3^{Cond_i}(t) + p_4^{Cond_i}(t)\right] \\ \dfrac{dp_2^{Cond_i}(t)}{dt} = 3\lambda^{Cond_i} p_1^{Cond_i}(t) - \left[2\lambda^{Cond_i} + \mu^{Cond_i}\right] p_2^{Cond_i}(t) \\ \dfrac{dp_3^{Con}(t)}{dt} = 2\lambda^{Cond_i} p_2^{Cond_i}(t) - \left[\lambda^{Cond_i} + \mu^{Cond_i}\right] p_3^{Cond_i}(t) \\ \dfrac{dp_4^{Cond_i}(t)}{dt} = \lambda^{Cond_i} p_3^{Cond_i}(t) - \mu^{Cond_i} p_4^{Cond_i}(t) \end{cases} \quad i = 1, 2$$

Initial conditions are:

$$p_1^{Cond_i}(0) = 1;\ p_2^{Cond_i}(0) = p_3^{Cond_i}(0) = p_4^{Cond_i}(0) = 0.$$

We used MATLAB® for the numerical solution of these systems of differential equations to obtain probabilities $p_1^{(i)}(t),\ p_2^{(i)}(t),\ (i =$ Comp$_1$, Comp$_2$, Pump$_1$, …, Pump$_5$, Ev$_1$, Ev$_2$, DTap) and $p_1^{Cond_i}(t),\ p_2^{Cond_i}(t),\ p_3^{Cond_i}(t),\ p_4^{Cond_i}(t),\ i = 1, 2$..

Therefore, for the elements of such systems, the output performance stochastic processes can be obtained in the following manner.

- For $i =$ Comp$_1$, Comp$_2$, Pump$_1$, Pump$_2$, Pump$_3$, Ev$_1$, Ev$_2$, SW$_1$, SW$_2$:

$$\begin{cases} \mathbf{g}^{(i)} = \left\{g_1^{(i)}, g_2^{(i)}\right\} = \{8, 0\}, \\ \mathbf{p}^{(i)}(t) = \left\{p_1^{(i)}(t), p_2^{(i)}(t)\right\}. \end{cases}$$

- For $i =$ Pump$_4$, Pump$_5$, Sw$_3$, DTap:

$$\begin{cases} \mathbf{g}^{(i)} = \left\{g_1^{(i)}, g_2^{(i)}\right\} = \{16, 0\}, \\ \mathbf{p}^{(i)}(t) = \left\{p_1^{(i)}(t), p_2^{(i)}(t)\right\}. \end{cases}$$

- For $i =$ Cond$_1$, Cond$_2$:

$$\begin{cases} \mathbf{g}^{(i)} = \left\{g_1^{(i)}, g_2^{(i)}, g_3^{(i)}, g_4^{(i)}\right\} = \{8, 6, 3, 0\}, \\ \mathbf{p}^{(i)}(t) = \left\{p_1^{(i)}(t), p_2^{(i)}(t), p_3^{(i)}(t), p_4^{(i)}(t)\right\}. \end{cases}$$

Sets $\mathbf{g}^{(i)}$, $\mathbf{p}^{(i)}(t)(i =$ Comp$_1$, Comp$_2$, Pump$_1$, …, Pump$_5$, Ev$_1$, Ev$_2$, DTap, Cond$_1$, Cond$_2$) define L_Z-transforms for each element as follows:

– Compressors (Comp$_i$, $i = 1,2$):

$$\begin{aligned} L_Z\left\{g^{Comp_i}(t)\right\} &= p_1^{Comp_i}(t) z^{g_1^{Comp_i}} + p_2^{Comp_i}(t) z^{g_2^{Comp_i}} \\ &= p_1^{Comp_i}(t) z^8 + p_2^{Comp_i}(t) z^0 \end{aligned} \tag{6.1}$$

– Pumps (Pump$_i$, $i = 1, 2, 3$):

$$\begin{aligned} Lz\{g^{Pump_i}(t)\} &= p_1^{Pump_i}(t)z^{g_1^{Pump_i}} + p_2^{Pump_i}(t)z^{g_2^{Pump_i}} \\ &= p_1^{Pump_i}(t)z^8 + p_2^{Pump_i}(t)z^0 \end{aligned} \tag{6.2}$$

- Pumps (Pump_i, $i = 4, 5$):

$$\begin{aligned} Lz\{g^{Pump_i}(t)\} &= p_1^{Pump_i}(t)z^{g_1^{Pump_i}} + p_2^{Pump_i}(t)z^{g_2^{Pump_i}} \\ &= p_1^{Pump_i}(t)z^{16} + p_2^{Pump_i}(t)z^0 \end{aligned} \tag{6.3}$$

- Evaporators (Ev_i, $i = 1, 2$):

$$Lz\{g^{Ev_i}(t)\} = p_1^{Ev_i}(t)z^{g_1^{Ev_i}} + p_2^{Ev_i}(t)z^{g_2^{Ev_i}} = p_1^{Ev_i}(t)z^8 + p_2^{Ev_i}(t)z^0 \tag{6.4}$$

- Switches (Sw_i, $i = 1, 2$):

$$Lz\{g^{Sw_i}(t)\} = p_1^{Sw_i}(t)z^{g_1^{Sw_i}} + p_2^{Sw_i}(t)z^{g_2^{Sw_i}} = p_1^{Sw_i}(t)z^8 + p_2^{Sw_i}(t)z^0 \tag{6.5}$$

- Switch (Sw_3):

$$Lz\{g^{Sw_3}(t)\} = p_1^{Sw_3}(t)z^{g_1^{Sw_3}} + p_2^{Sw_3}(t)z^{g_2^{Sw_3}} = p_1^{Sw_3}(t)z^{16} + p_2^{Sw_3}(t)z^0 \tag{6.6}$$

- Differential tap (Dtap):

$$\begin{aligned} Lz\{g^{DTap}(t)\} &= p_1^{DTap}(t)z^{g_1^{DTap}} + p_2^{DTap}(t)z^{g_2^{DTap}} \\ &= p_1^{DTap}(t)z^{16} + p_2^{DTap}(t)z^0 \end{aligned} \tag{6.7}$$

- Condensers (Cond_i, $i = 1, 2$):

$$\begin{aligned} Lz\{g^{Cond_i}(t)\} &= p_1^{Cond_i}(t)z^{g_1^{Cond_i}} + p_2^{Cond_i}(t)z^{g_2^{Cond_i}} \\ &\quad + p_3^{Cond_i}(t)z^{g_3^{Cond_i}} + p_4^{Cond_i}(t)z^{Cond_i} \\ &= p_1^{Cond_i}(t)z^8 + p_2^{Cond_i}(t)z^6 + p_3^{Cond_i}(t)z^3 + p_4^{Cond_i}(t)z^0 \end{aligned} \tag{6.8}$$

6.2 The L_Z-Transform Application for Calculation of the Availability and Performability Indices of the Cold Water Supply System

According to Fig. 6.2 and using the recursive procedure [6], the multi-state model of the cold water supply system (CWSS) can be presented using the L_Z-transform as follows:

$$
\begin{aligned}
L_Z\left\{G^{Ch_1}(t)\right\} &= \Omega_{f_{ser}}\left(L_Z\left\{G^{Comp_1}(t)\right\}, L_Z\left\{G^{Cond_1}(t)\right\}, L_Z\left\{G^{Ev_1}(t)\right\}\right);\\
L_Z\left\{G^{PS_1}(t)\right\} &= \Omega_{f_{ser}}\left(L_Z\left\{G^{Pump_1}(t)\right\}, L_Z\left\{G^{Sw_1}(t)\right\}\right);\\
L_Z\left\{G^{ChP_1}(t)\right\} &= \Omega_{f_{ser}}\left(L_Z\left\{G^{Ch_1}(t)\right\}, L_Z\left\{G^{PS_1}(t)\right\}\right);\\
L_Z\left\{G^{Ch_2}(t)\right\} &= \Omega_{f_{ser}}\left(L_Z\left\{G^{Comp_2}(t)\right\}, L_Z\left\{G^{Cond_2}(t)\right\}, L_Z\left\{G^{Ev_2}(t)\right\}\right);\\
L_Z\left\{G^{PS_2}(t)\right\} &= \Omega_{f_{ser}}\left(\Omega_{f_{par}}\left(L_Z\left\{G^{Pump_2}(t)\right\}, L_Z\left\{G^{Pump_3}(t)\right\}\right), L_Z\left\{G^{Sw_2}(t)\right\}\right);\\
L_Z\left\{G^{ChP_2}(t)\right\} &= \Omega_{f_{ser}}\left(L_Z\left\{G^{Ch_2}(t)\right\}, L_Z\left\{G^{PS_2}(t)\right\}\right);\\
L_Z\left\{G^{ChP}(t)\right\} &= \Omega_{f_{par}}\left(L_Z\left\{G^{ChP_1}(t)\right\}, L_Z\left\{G^{ChP_2}(t)\right\}\right);\\
L_Z\left\{G^{PS_3}(t)\right\} &= \Omega_{f_{ser}}\left(\Omega_{f_{1-out-of-2}}\left(L_Z\left\{G^{Pump_4}(t)\right\}, L_Z\left\{G^{Pump_5}(t)\right\}\right), L_Z\left\{G^{Sw_3}(t)\right\}\right);\\
L_Z\left\{G^{CWSS}(t)\right\} &= \Omega_{f_{ser}}\left(L_Z\left\{G^{ChP}(t)\right\}, L_Z\left\{g^{DTap}(t)\right\}, L_Z\left\{G^{PS_3}(t)\right\}\right).
\end{aligned}
\tag{6.9}
$$

Let us calculate the L_Z-transforms according to expressions (6.9).

Using the composition operator $\Omega_{f_{ser}}$, we obtain the L_Z-transform $L_Z\left\{G^{Ch_i}(t)\right\}$ for the first and the second chillers, where powers of z are calculated as the minimum of powers of the corresponding terms:

$$
\begin{aligned}
L_Z\{G^{Ch_i}(t)\} &= \Omega_{f_{ser}}\left(L_Z\{G^{Comp_i}(t)\}, L_Z\{G^{Cond_i}(t)\}, L_Z\{G^{Ev_i}(t)\}\right)\\
&= \Omega_{f_{ser}}\Big(p_1^{Comp_i}(t)z^8 + p_2^{Comp_i}(t)z^0,\ p_1^{Cond_i}(t)z^8 + p_2^{Cond_i}(t)z^6\\
&\quad + p_3^{Cond_i}(t)z^3 + p_4^{Cond_i}(t)z^0,\ \ p_1^{Ev_i}(t)z^8 + p_2^{Ev_i}(t)z^0\Big)\\
&= P_1^{Ch_i}(t)z^8 + P_2^{Ch_i}(t)z^6 + P_3^{Ch_i}(t)z^3 + P_4^{Ch_i}(t)z^0,\ i = 1, 2
\end{aligned}
\tag{6.10}
$$

where

$$
\begin{aligned}
P_1^{Ch_i}(t) &= p_1^{Comp_i}(t)\cdot p_1^{Cond_i}(t)\cdot p_1^{Ev_i}(t);\\
P_2^{Ch_i}(t) &= p_1^{Comp_i}(t)\cdot p_2^{Cond_i}(t)\cdot p_1^{Ev_i}(t);\\
P_3^{Ch_i}(t) &= p_1^{Comp_i}(t)\cdot p_3^{Cond_i}(t)\cdot p_1^{Ev_i}(t);\\
P_4^{Ch_i}(t) &= \left(p_1^{Comp_i}(t)\cdot p_4^{Cond_i}(t) + p_2^{Comp_i}(t)\right)\cdot p_1^{Ev_i}(t) + p_2^{Ev_i}(t);
\end{aligned}
$$

Using the composition operator Ω_{fser}, we obtain the L_Z-transform $L_Z\{G^{PS_1}(t)\}$ for the connected in series the first pump and the first switch, where the powers of z are calculated as the minimum of powers of the corresponding terms:

$$\begin{aligned} L_Z\{G^{PS_1}(t)\} &= \Omega_{f_{ser}}\left(L_Z\{G^{Pump_1}(t)\}, L_Z\{G^{Sw_1}(t)\}\right) \\ &= \Omega_{f_{ser}}\left(p_1^{Pump_1}(t)z^8 + p_2^{Pump_1}(t)z^0, p_1^{Sw_1}(t)z^8 + p_2^{Sw_1}(t)z^0\right. \\ &= P_1^{PS_1}(t)z^8 + P_2^{PS_1}(t)z^0 \end{aligned} \tag{6.11}$$

where

$$\begin{aligned} P_1^{PS_1}(t) &= p_1^{Pump_1}(t) \cdot p_1^{Sw_1}(t); \\ P_1^{PS_1}(t) &= p_1^{Pump_1}(t) \cdot p_2^{Sw_1}(t) + p_2^{Pump_1}(t); \end{aligned}$$

Using the composition operator Ω_{fser}, we obtain the L_Z-transform $L_Z\{G^{ChP_1}(t)\}$ for the connected in series the first chiller, the first pump and the first switch, where the powers of z are calculated as the minimum of powers of the corresponding terms:

$$\begin{aligned} L_Z\{G^{ChP_1}(t)\} &= \Omega_{f_{ser}}\left(L_Z\{G^{Ch_1}(t)\}, L_Z\{G^{PS_1}(t)\}\right) \\ &= \Omega_{f_{ser}}\left(P_1^{Ch_1}(t)z^8 + P_2^{Ch_1}(t)z^6 + P_3^{Ch_1}(t)z^3\right. \\ &\quad \left. + P_4^{Ch_1}(t)z^0, P_1^{PS_1}(t)z^8 + P_2^{PS_1}(t)z^0\right) \\ &= P_1^{ChP_1}(t)z^8 + P_2^{ChP_1}(t)z^6 f + P_3^{ChP_1}(t)z^3 + P_4^{ChP_1}(t)z^0 \end{aligned} \tag{6.12}$$

where

$$\begin{aligned} P_1^{ChP_1}(t) &= P_1^{Ch_1}(t) \cdot P_1^{PS_1}(t); \\ P_2^{ChP_1}(t) &= P_2^{Ch_1}(t) \cdot P_1^{PS_1}(t); \\ P_3^{ChP_1}(t) &= P_3^{Ch_1}(t) \cdot P_1^{PS_1}(t); \\ P_4^{ChP_1}(t) &= P_4^{Ch_1}(t) \cdot P_1^{PS_1}(t) + P_2^{PS_1}(t); \end{aligned}$$

Using the composition operators Ω_{fser} and Ω_{fpar}, we obtain the $L_Z$transform $L_Z\{G^{Ch_2}(t)\}$ for the second and the third pumps and the second switch:

$$\begin{aligned} L_Z\left\{G^{Ch_2}(t)\right\} &= \Omega_{f_{ser}}\left(\Omega_{f_{par}}\left(L_Z\left\{G^{Pump_2}(t)\right\}, L_Z\left\{G^{Pump_3}(t)\right\}\right),\right. \\ &\quad \left. L_Z\left\{G^{Sw_2}(t)\right\}\right) \\ &= \Omega_{f_{ser}}\left(\Omega_{f_{par}}\left(p_1^{Pump_2}(t)z^8 + p_2^{Pump_2}(t)z^0,\right.\right. \\ &\quad \left. * p_1^{Pump_3}(t)z^8 + p_2^{Pump_3}(t)z^0\right), \\ &\quad \left. p_1^{Sw_2}(t)z^8 + p_2^{Sw_2}(t)z^0\right) \\ &= P_1^{Ch_2}(t)z^8 + P_2^{Ch_2}(t)z^0 \end{aligned} \tag{6.13}$$

where

$$P_1^{Ch_2}(t) = p_1^{Sw_2}(t) \cdot \left(p_1^{Pump_2}(t) \cdot p_2^{Pump_3}(t) + p_1^{Pump_3}(t)\right);$$
$$P_2^{Ch_2}(t) = p_2^{Pump_2}(t) \cdot p_2^{Pump_3}(t) + p_2^{Sw_2}(t) \cdot \left(p_2^{Pump_2}(t) \cdot p_1^{Pump_3}(t) + p_1^{Pump_2}(t)\right);$$

Using the composition operator Ω_{fser}, we obtain the L_Z-transform $L_Z\{G^{ChP_2}(t)\}$ for the connected in series the second chiller, the second and third pumps and the second switch:

$$\begin{aligned} L_Z\{G^{ChP_2}(t)\} &= \Omega_{f_{ser}}\left(L_Z\{G^{Ch_2}(t)\}, L_Z\{G^{PS_2}(t)\}\right) \\ &= \Omega_{f_{ser}}\Big(P_1^{Ch_2}(t)z^8 + P_2^{Ch_2}(t)z^6 + P_3^{Ch_2}(t)z^3 + P_4^{Ch_2}(t)z^0, \\ &\quad P_1^{PS_2}(t)z^8 + P_2^{PS_2}(t)z^0\Big) \\ &= P_1^{ChP_2}(t)z^8 + P_2^{ChP_2}(t)z^6 + P_3^{ChP_2}(t)z^3 + P_4^{ChP_2}(t)z^0 \end{aligned} \tag{6.14}$$

where

$$\begin{aligned} P_1^{ChP_2}(t) &= P_1^{Ch_2}(t) \cdot P_1^{PS_2}(t); \\ P_2^{ChP_2}(t) &= P_2^{Ch_2}(t) \cdot P_1^{PS_2}(t); \\ P_3^{ChP_2}(t) &= P_3^{Ch_2}(t) \cdot P_1^{PS_2}(t); \\ P_4^{ChP_2}(t) &= P_4^{Ch_2}(t) \cdot P_1^{PS_2}(t) + P_2^{PS_2}(t); \end{aligned}$$

Using the composition operator Ω_{fpar}, we obtain the L_Z-transform $L_Z\{G^{ChP}(t)\}$ for the connected in parallel two chiller, pumps and switches sub-systems:

$$\begin{aligned} L_Z\left\{G^{ChP}(t)\right\} &= \Omega_{f_{par}}\left(L_Z\left\{G^{ChP_2}(t)\right\}, L_Z\left\{G^{ChP_2}(t)\right\}\right) \\ &= \Omega_{f_{ser}}\Big(P_1^{ChP_1}(t)z^8 + P_2^{ChP_1}(t)z^6 + P_3^{ChP_1}(t)z^3 + P_4^{ChP_1}(t)z^0, \\ &\quad P_1^{ChP_2}(t)z^8 + P_2^{ChP_2}(t)z^6 + P_3^{ChP_2}(t)z^3 + P_4^{ChP_2}(t)z^0\Big) \\ &= P_1^{ChP}(t)z^{16} + P_2^{ChP}(t)z^{14} + P_3^{ChP}(t)z^{12} + P_4^{ChP}(t)z^{11} + P_5^{ChP}(t)z^9 \\ &\quad + P_6^{ChP}(t)z^8 + P_7^{ChP}(t)z^6 + P_8^{ChP}(t)z^3 + P_9^{ChP}(t)z^0 \end{aligned} \tag{6.15}$$

where

$$\begin{aligned} P_1^{ChP}(t) &= P_1^{ChP_1}(t) \cdot P_1^{ChP_2}(t); \\ P_2^{ChP}(t) &= P_1^{ChP_1}(t) \cdot P_2^{ChP_2}(t) + P_1^{ChP_2}(t) \cdot P_2^{ChP_1}(t); \\ P_3^{ChP}(t) &= P_2^{ChP_1}(t) \cdot P_2^{ChP_2}(t); \end{aligned}$$

$$P_4^{ChP}(t) = P_1^{ChP_1}(t) \cdot P_3^{ChP_2}(t) + P_3^{ChP_1}(t) \cdot P_1^{ChP_1}(t);$$
$$P_5^{ChP}(t) = P_2^{ChP_1}(t) \cdot P_3^{ChP_2}(t) + P_3^{ChP_1}(t) \cdot P_2^{ChP_2}(t);$$
$$P_6^{ChP}(t) = P_1^{ChP_1}(t) \cdot P_4^{ChP_2}(t) + P_4^{ChP_1}(t) \cdot P_1^{ChP_2}(t);$$
$$P_7^{ChP}(t) = P_2^{ChP_1}(t) \cdot P_4^{ChP_2}(t) + P_3^{ChP_1}(t) \cdot P_3^{ChP_2}(t) + P_4^{ChP_1}(t) \cdot P_2^{ChP_2}(t);$$
$$P_8^{ChP}(t) = P_3^{ChP_1}(t) \cdot P_4^{ChP_2}(t) + P_4^{ChP_1}(t) \cdot P_3^{ChP_2}(t);$$
$$P_9^{ChP}(t) = P_4^{ChP_1}(t) \cdot P_4^{ChP_2}(t);$$

Using the composition operators Ω_{fser} and $\Omega_{f1-out-of-2}$, we obtain the L_Z-transform $L_Z\{G^{PS_3}(t)\}$ for the forth and the fifth pumps and the third switch:

$$\begin{aligned} L_Z\{G^{PS_3}(t)\} &= \Omega_{f_{ser}}\left(\Omega_{f1-out-of-2}\left(L_Z\{G^{Pump_4}(t)\}, L_Z\{G^{Pump_5}(t)\}\right), L_Z\{G^{Sw_3}(t)\}\right) \\ &= \Omega_{f_{ser}}\left(\Omega_{f1-out-of-2}\left(p_1^{Pump_4}(t)z^{16} + p_2^{Pump_4}(t)z^0, p_1^{Pump_5}(t)z^{16} + p_2^{Pump_5}(t)z^0\right),\right. \\ &\quad \left. p_1^{Sw_3}(t)z^{16} + p_2^{Sw_3}(t)z^0\right) = P_1^{PS_3}(t)z^{16} + P_2^{PS_3}(t)z^0 \end{aligned} \tag{6.16}$$

where

$$P_1^{PS_3}(t) = p_1^{Sw_3}(t) \cdot \left(p_2^{Pump_4}(t) \cdot p_1^{Pump_5}(t) + p_1^{Pump_4}(t)\right);$$
$$P_2^{PS_3}(t) = p_1^{Sw_3}(t) \cdot p_2^{Pump_4}(t) \cdot p_2^{Pump_5}(t) + p_2^{Sw_3}(t);$$

Using the composition operator Ω_{fser} for chillers sub-system, differential tap and for the forth and the fifth pumps and the third switch subsystem, we obtain the L_Z-transforms $L_Z\{G^{CWSS}(t)\}$ for the entire cold water supply system:

$$\begin{aligned} L_Z\{G^{CWSS}(t)\} &= \Omega_{f_{ser}}\left(L_Z\{G^{ChP}(t)\}, L_Z\{G^{DTap}(t)\}, L_Z\{G^{PS_3}(t)\}\right) \\ &= P_1^{CWSS}(t)z^{16} + P_2^{CWSS}(t)z^{14} + P_3^{CWSS}(t)z^{12} + P_4^{CWSS}(t)z^{11} \\ &\quad + P_5^{CWSS}(t)z^9 + P_6^{CWSS}(t)z^8 + P_7^{CWSS}(t)z^6 \\ &\quad + P_8^{CWSS}(t)z^3 + P_9^{CWSS}(t)z^0 \end{aligned} \tag{6.17}$$

where

$$P_1^{CWSS}(t) = P_1^{ChP}(t) \cdot P_1^{DTap}(t) \cdot P_1^{PS_3}(t);$$
$$P_2^{CWSS}(t) = P_2^{ChP}(t) \cdot P_1^{DTap}(t) \cdot P_1^{PS_3}(t);$$
$$P_3^{CWSS}(t) = P_3^{ChP}(t) \cdot P_1^{DTap}(t) \cdot P_1^{PS_3}(t);$$
$$P_4^{CWSS}(t) = P_4^{ChP}(t) \cdot P_1^{DTap}(t) \cdot P_1^{PS_3}(t);$$
$$P_5^{CWSS}(t) = P_5^{ChP}(t) \cdot P_1^{DTap}(t) \cdot P_1^{PS_3}(t);$$
$$P_6^{CWSS}(t) = P_6^{ChP}(t) \cdot P_1^{DTap}(t) \cdot P_1^{PS_3}(t);$$
$$P_7^{CWSS}(t) = P_7^{ChP}(t) \cdot P_1^{DTap}(t) \cdot P_1^{PS_3}(t);$$

$$P_8^{CWSS}(t) = P_8^{ChP}(t) \cdot P_1^{DTap}(t) \cdot P_1^{PS_3}(t);$$
$$P_9^{CWSS}(t) = P_9^{ChP}(t) \cdot P_1^{DTap}(t) \cdot P_1^{PS_3}(t) + P_2^{PS_3}(t) + P_2^{DTap}(t) \cdot P_1^{PS_3}(t);$$

6.3 Calculation of the Reliability and Performability Indices of the Cold Water Supply System

Using expression (3.28) from the Chap. 3, the instantaneous availability for constant demand levels w may be presented as follows:

- High demand level ($w \geq 16$)

$$A_{w\geq 16}^{CWSS}(t) = \sum_{g_i^{CWSS}\geq 16} P_i^{CWSS}(t) = P_1^{CWSS}(t) \tag{6.18}$$

- Low demand level ($w \geq 10$)

$$A_{w\geq 10}^{CWSS}(t) = \sum_{g_i^{CWSS}\geq 10} P_i^{CWSS}(t) = P_1^{CWSS}(t) + P_2^{CWSS}(t) + P_3^{CWSS}(t) + P_4^{CWSS}(t). \tag{6.19}$$

The mean instantaneous performance (3.29) from the Chap. 3 for CWSS can be obtained in the following manner:

$$\begin{aligned} E^{CWSS}(t) &= \sum_{g_i^{DGSW}>0} g_i^{CWSS} P_i^{CWSS}(t) = \sum_{i=1}^{9} g_i^{CWSS} P_i^{CWSS}(t) \\ &= 16 \cdot P_1^{CWSS}(t) + 14 \cdot P_2^{CWSS}(t) + 12 \cdot P_3^{CWSS}(t) + 11 \cdot P_4^{CWSS}(t) \\ &\quad + 9 \cdot P_5^{CWSS}(t) + 8 \cdot P_6^{CWSS}(t) + 6 \cdot P_7^{CWSS}(t) + 3 \cdot P_8^{CWSS}(t). \end{aligned} \tag{6.20}$$

Calculations were performed using the following failure and repair rates.

The failure rates of each compressor are $\lambda_1^{Com} = \lambda_2^{Com} = 0.25$ year^{-1}. The repair rates are $\mu_1^{Com} = \mu_2^{Com} = 180$ year^{-1}. The failure rates of each condenser are $\lambda_1^{Con} = \lambda_2^{Con} = 1$ year^{-1}. The repair rates are $\mu_1^{Con} = \mu_2^{Con} = 365$ year^{-1}. The failure rates of each evaporator are $\lambda_1^{EV} = \lambda_2^{EV} = 0.1$ year^{-1}. The repair rates are $\mu_1^{EV} = \mu_2^{EV} = 52$ year^{-1}.

The failure rates of the pumps are $\lambda_1^P = \lambda_2^P = \lambda_3^P = \lambda_4^P = 0.5$ year^{-1}. The repair rates are $\mu_1^P = \mu_2^P = \mu_3^P = \mu_4^P = 365$ year^{-1} .

The failure rate of switches are $\lambda_1^{SW} = \lambda_2^{SW} = \lambda_3^{SW} = 0.25$ year^{-1}. The repair rates are $\mu_1^{SW} = \mu_2^{SW} = \mu_3^{SW} = 365$ year^{-1} .

The failure rate of the differential tap is $\lambda^{Dtap} = 0.1 + t^2$ year^{-1}. The repair rates are $\mu^{EV} = 52$ year^{-1}. As one can see, the failure rate of the differential tap is increasing functions of time, this element possessing the aging property.

The graphs of CWSS instantaneous availability for different demand levels and mean instantaneous performance are shown in Figs. 6.5 and 6.6.

Availability calculations of MSS cold water supply system are presented on Fig. 6.5. The curves on this figure show that the availability of an aging system decreases with time. From the Fig. 6.5, one can see that during the first few days of operation of the system, there is a very sharp decrease in the availability of the system, and then availability decreases moderately, subject to the aging process of the differential tap. Fig. 6.5 shows that the system provides the ASHRAE [1] availability requirement (greater than 0.92) only during the first two years of exploitation. Further exploitation impossible without repair or replacement of the differential tap.

From the Fig. 6.6, one can see that during two years of exploitation, because the aging property of the differential tap performance of the system reduces on 8.5%. This result also supports the need of repair or replacement of the differential tap after two years of exploitation.

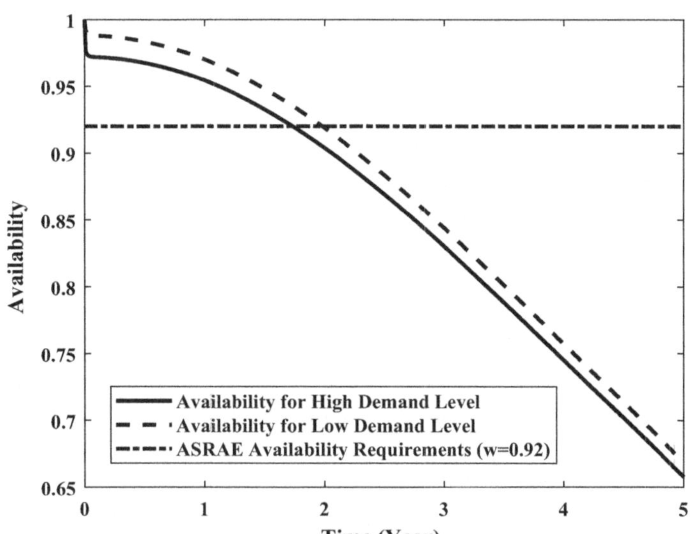

Fig. 6.5 The cold water supply system instantaneous availability for different demand levels and ASHRAE requirements

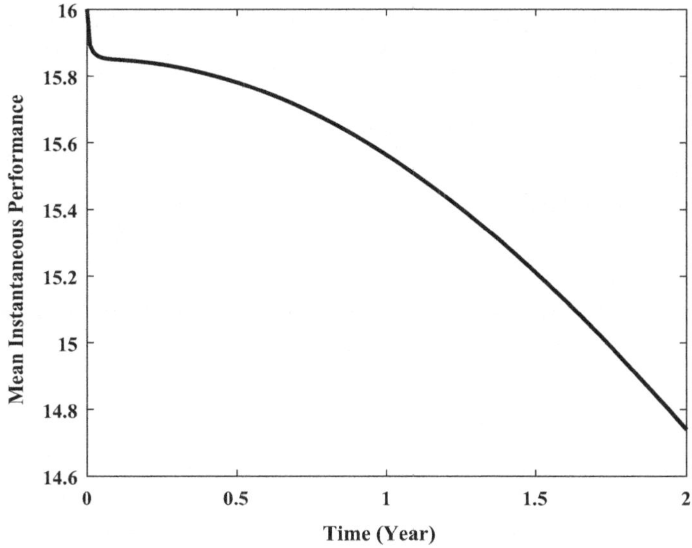

Fig. 6.6 The Cold Water Supply System mean instantaneous performance

6.4 Summary

In this chapter, we applied the L_Z-transform to a practical problem of availability and performability calculation for MSS with aging. The cold water supply system for a raw material manufacturing for plastics industry was considered.

The calculated results show decreasing the availability and performability of the cold water supply system because the aging property of the differential tap. The availability curves show that the system provides the ASHRAE availability requirement only during the first two years, and repair or replacement of the differential tap after two years period of exploitation is needed.

References

1. Handbook ASHRAE (2014) Refrigeration. American society of heating refrigeratinsg and air-conditioning engineers, Atlanta
2. Frenkel I, Khvatskin L (2012) Reliability decision for supermarket refrigiration system by using combine stochastic process and universal generating function method: Case Study. In: Lisnianski A, Frenkel I (eds) Recent advances in system reliability. Signatures, multi-state systems and statistical inference. Springer, London pp 97–112
3. Frenkel I, Lisnianski A, Khvatskin L (2014) On Lz-transform application for availability assessment of aging multi-state water cooling system for medical equipment. In: Frenkel I, Karagrigoriou A, Lisnianski A, Kleyner A (eds) Applied reliability engineering and risk analysis: probabilistic models and statistical inference. Wiley, Chichester, pp 59–77

4. Lein P, Frenkel I, Khvatskin L (2016) On availability determination for mss cold water supply system by lz-transform: case study. In: Frenkel I, Lisnianski A (eds) Proceedings of the second international symposium on stochastic models in reliability engineering, life science and operations management (SMRLO'16), Beer Sheva, Israel, February 15–18, 2016, pp 365–374. IEEE CPS, 978-1-4673-9941-8/16. https://doi.org/10.1109/smrlo.2016.65
5. Lisnianski A (2012) *Lz*-transform for a Discrete-state continuous-time markov process and its application to multi-state system reliability. In: Lisnianski A, Frenkel I (eds) Recent advances in system reliability. Signatures, multi-state systems and statistical inference. Springer, London, pp 79–95
6. Lisnianski A, Frenkel I, Ding Y (2010) Multi-state system reliability analysis and optimization for engineers and industrial managers. Springer, London
7. Natvig B (2011) Multistate systems reliability. Theory with applications. Wiley, New York

Chapter 7
Sensitivity Evaluation for an Aging Multi-state System

Investigation of an impact of changing different failure or repair rates in different elements in multi-state system (MSS) is often important for practical reliability engineering. Based on this reliability researcher or engineer can make appropriate decision for MSS reliability improvement. A most common approach for sensitivity analysis is changing one factor at a time (OAT) [3, 5], to see what effect this produces on the output. OAT involves the following steps: (a) moving one input variable, keeping others at their baseline (nominal) values; (b) returning the variable to its nominal value, and repeating for each of the other inputs the same procedure. Sensitivity may then be measured by monitoring changes in the output by partial derivatives.

When we are dealing with MSS, the most important measures are a system availability, mean MSS output performance, mean output performance deficiency, mean accumulated performance deficiency.

For aging MSS such parameters as availability and mean output performance are usually decreasing over time t, and accumulated performance deficiency is increasing.

In practice reliability engineer often is interested in estimating of impact of changing of different system parameters such as failure and repair rates of different MSS components to the entire system parameters [4].

In this chapter we present the method for aging MSS sensitivity analysis, which is based on L_Z-transform application [2, 3]. It will be shown that by using the method one can avoid a building of Markov model for entire MSS and solution of corresponding system with great number of differential equations. Instead of this one can solve only simple models for the system's elements and then use a simple algebra.

A. Lisnianski et al., *Modern Dynamic Reliability Analysis for Multi-state Systems*, Springer Series in Reliability Engineering,
https://doi.org/10.1007/978-3-030-52488-3_7

7.1 Sensitivity Analysis for Aging MSS by Using L_Z-Transform Method

If L_Z-transform of output stochastic process is known ((3.27–3.32) from the Chap. 3), then important reliability measures such as availability, mean expected performance, expected accumulated performance deficiency, etc., can be found for aging MSS [3].

For required demand level w, MSS availability at instant $t \geq 0$ is defined by the following expression:

$$A_w(t) = \sum_{g_i \geq w} p_i(t) \tag{7.1}$$

For non-aging MSS all transition intensities are constants. The sensitivity measure of availability corresponding to failure/repair rates changing in this case is the following

$$S_{a_{lm}^{(j)}}^{A_w} = \frac{\partial A_w}{\partial a_{lm}^{(j)}} \tag{7.2}$$

Remind that $a_{lm}^{(j)}$ is transition intensity for the transition from state l to state m in the multi-state component j.

For aging MSS availability $A_w(t)$ is decreasing function of time. Therefore, sensitivity measure of availability corresponding to failure/repair rates will be also a function of time

$$S_{a_{lm}^{(j)}}^{A_w}(t) = \frac{\partial A_w(t)}{\partial a_{lm}^{(j)}(t)} \cdot \frac{\partial a_{lm}^{(j)}(t)}{\partial t} \tag{7.3}$$

Because of time-dependence, straightforward usage of this measure for aging system is often not suitable in practical reliability engineering.

Note, that in aging system, consisting of arbitrary number of components, only minority (usually one or two) of components are aging. Majority of components are non-aging with constant failure rates. The method presented in the paper is appropriate for this majority.

In practice, it is often important to know how much time remains until the instant, when the availability will down crossing some specified level A_0. We designate this time as $T_{A_w \geq A_0}$. Usually, when such down crossing occurs, some extended maintenance actions should be performed. For example, it may be replacement of an aging component or it's overall.

Keeping in mind this fact, we shall say that ***time*** $T_{A_w \geq A_0}$ ***determines aging MSS life time (up to extended maintenance or replacement) with respect to availability restriction.***

Typical problem of MSS sensitivity analysis in this context is the following: what is an impact of decreasing or increasing of component's j transition intensity (failure or repair) rate $a_{lm}^{(j)}$ to an aging system life time $T_{A_w \geq A_0}$?

Then a ***sensitivity measure for aging MSS life time with respect to availability restriction*** can be determined as follows:

$$S_{a_{lm}^{(j)}}^{A_w} = \frac{\partial T_{A_w \geq A_0}}{\partial a_{lm}^{(j)}} \tag{7.4}$$

In some cases, mean instantaneous MSS output performance $\bar{G}(t)$ may be a crucial factor that determines entire system life time. We designate

$$\bar{G}(t) = E\{G(t)\} \tag{7.5}$$

where E is an expectation symbol, $G(t)$ is MSS output performance stochastic process. $\bar{G}(t)$ can be obtained based on L_Z-transform ((3.28) from the Chap. 3)

$$\bar{G}(t) = \sum_{i=1}^{K} g_i p_i(t) \tag{7.6}$$

We designate $T_{\bar{G} \geq G_0}$ the time up to the instant, when mean value will down crossing some preliminary specified level G_0.

Then a ***sensitivity measure for aging MSS life time with respect to mean output performance restriction*** can be determined for an aging system

$$S_{a_{lm}^{(j)}}^{\bar{G}} = \frac{\partial T_{\bar{G} \geq G_0}}{\partial a_{lm}^{(j)}} \tag{7.7}$$

In some other cases, an accumulated performance deficiency may be restricted from technical reasons. Instantaneous performance deficiency $d_w(t)$ characterizes output system performance $G(t)$ deviation from demand w, when the demand w is not met. Therefore, we have

$$d_w(t) = \max\{w - G(t), 0\} \tag{7.8}$$

By using ((3.31) from the Chap. 3) the expected instantaneous performance deficiency at instant t can be obtained

$$d_w^{(E)}(t) = \sum_{i=1}^{K} p_i(t) \max\{w - g_i, 0\} \tag{7.9}$$

The expected accumulated performance deficiency during time interval [0, t] is defined as follows:

$$D_w(t) = \int_0^t d_w^{(E)}(x)dx \tag{7.10}$$

For example, in power system D_w is the energy not supplied to consumers during the time interval [0, t].

Time $T_{D_w \leq D_0}$ determines the time remaining until the instant, when the accumulated performance deficiency will be greater than previously specified level D_0.

Then a ***sensitivity measure for aging MSS life time with respect to accumulated performance deficiency restriction*** can be determined for aging system by the following manner

$$S_{a_{lm}^{(j)}}^{D_w} = \frac{\partial T_{D_w \leq D_0}}{\partial a_{lm}^{(j)}} \tag{7.11}$$

The sensitivity measures, which were defined above, are mostly useful for aging MSS in practical reliability engineering.

7.2 Multi-state Models of the Water Cooling System for Magnetic Resonance Inspection (MRI) Plant

7.2.1 System's Description

As numerical example, we consider a water cooling system for magnetic resonance inspection (MRI) plant that is widely used in hospitals.

If the water cooling system fails, the aftermath may be very serious: evaporation of helium is strengthened and the helium stream escapes through special tubes with the risk of destroying not only the MRI plant, but also neighboring buildings. Because of these reasons, required availability level should be enough high. In according to [1], required availability should not be less than $A_0 = 0.97$.

The system consists of chillers' subsystem, heat exchanger and pumps' subsystem. The corresponding block diagram is presented in Fig. 7.1. Inside of each block depicted its states-transitions diagram and performance levels are presented near corresponding states. For simplification, we omit a multiplier 10^4 and present cooling capacities (performances) in 10^4 BTU per hour units.

The maximal performance (cooling capacity) of entire system, when all its elements are perfect, is 24×10^4 BTU/h.

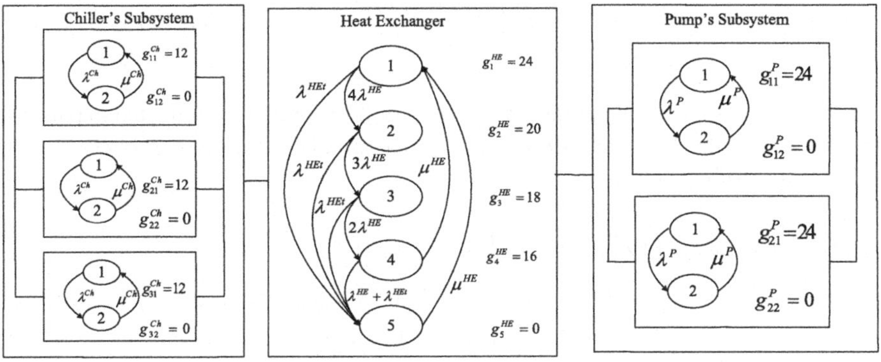

Fig. 7.1 Structure of the water cooling system for magnetic resonance inspection

7.2.2 Description of Sub-systems and System's Elements

7.2.2.1 Chiller Sub-system

The chiller sub-system consists of three connected in parallel packaged chillers. Each chiller have no possibility for cooling regulation, so their performance can be allowed in one of two states: "full capacity" or "zero capacity". According to the equipment type, the cooling capacity of each chiller is 24×10^4 BTU/h.

According to L_Z-transform method, in order to calculate the probabilities for each states, we build the corresponding systems of differential equations [6] for each chiller:

$$\begin{cases} \dfrac{dp_{i1}^{Ch}(t)}{dt} = -\lambda_i p_{i1}^{Ch}(t) + \mu_i p_{i2}^{Ch}(t), \\ \dfrac{dp_{i2}^{Ch}(t)}{dt} = \lambda_i p_{i1}^{Ch}(t) - \mu_i p_{i2}^{Ch}(t). \end{cases} \quad i = 1, 2, 3 \tag{7.12}$$

Initial conditions are: $p_{i1}^{Ch}(0) = 1;\ p_{i2}^{Ch}(0) = 0$.

We used MATLAB® for the numerical solution of these systems of differential equations to obtain probabilities $p_{i1}^{Ch}(t)$, $p_{i2}^{Ch}(t)$, $(i = 1, 2, 3)$.

Therefore, for the elements of chiller sub-systems, the output performance stochastic processes can be obtained in the following manner:

$$\begin{cases} \mathbf{g}_i^{Ch} = \{g_{i1}^{Ch}, g_{i2}^{Ch}\} = \{12, 0\}, \\ \mathbf{p}_i^{Ch}(t) = \{p_{i1}^{Ch}(t), p_{i2}^{Ch}(t)\}. \end{cases}$$

Sets $\mathbf{g}_i^{Ch}$, $\mathbf{p}_i^{Ch}(t)$ $(i = 1, 2, 3)$ define L_Z-transforms for each chillers as follows:

$$L_Z\{g_i^{Ch}(t)\} = p_{i1}^{Ch}(t)z^{g_{i1}^{Ch}} + p_{i2}^{Ch}(t)z^{g_{i2}^{Ch}} = p_{i1}^{Ch}(t)z^{12} + p_{i2}^{Ch}(t)z^{0} \quad i = 1, 2, 3 \tag{7.13}$$

Using the composition operator Ω_{fpar} [3] for three chillers, connected in parallel, one can obtain the L_Z-transform for output performance stochastic process $G^{Ch}(t)$ of the chiller sub-system

$$\begin{aligned} L_Z\left\{G^{Ch}(t)\right\} &= \Omega_{fpar}\left(L_Z\left\{g_1^{Ch}(t)\right\}, L_Z\left\{g_2^{Ch}(t)\right\}, L_Z\left\{g_3^{Ch}(t)\right\}\right) \\ &= \Omega_{fpar}\left(p_{11}^{Ch}(t)z^{12} + p_{12}^{Ch}(t)z^{0}, p_{21}^{Ch}(t)z^{12} + p_{22}^{Ch}(t)z^{0}, p_{31}^{Ch}(t)z^{12} + p_{32}^{Ch}(t)z^{0}\right) \\ &= P_1^{Ch}(t)z^{36} + P_2^{Ch}(t)z^{24} + P_3^{Ch}(t)z^{12} + P_4^{Ch}(t)z^{0} \end{aligned} \tag{7.14}$$

where

$$\begin{aligned} P_1^{Ch}(t) &= p_{11}^{Ch}(t)p_{21}^{Ch}(t)p_{31}^{Ch}(t); \\ P_2^{Ch}(t) &= p_{11}^{Ch}(t)p_{21}^{Ch}(t)p_{32}^{Ch}(t) + p_{11}^{Ch}(t)p_{22}^{Ch}(t)p_{31}^{Ch}(t) + p_{12}^{Ch}(t)p_{21}^{Ch}(t)p_{31}^{Ch}(t); \\ P_3^{Ch}(t) &= p_{11}^{Ch}(t)p_{22}^{Ch}(t)p_{32}^{Ch}(t) + p_{12}^{Ch}(t)p_{21}^{Ch}(t)p_{32}^{Ch}(t) + p_{12}^{Ch}(t)p_{22}^{Ch}(t)p_{31}^{Ch}(t); \\ P_4^{Ch}(t) &= p_{12}^{Ch}(t)p_{22}^{Ch}(t)p_{32}^{Ch}(t). \end{aligned}$$

7.2.2.2 Heat Exchanger

The heat exchanger is a device with 5 states of performance degradation: a fully operational state with a capacity of 24×10^4 BTU/h, states of partial failure corresponding to capacities $22 \times 10^4, 20 \times 10^4, 16 \times 10^4$ of BTU per hour and a total failure corresponding to a capacity of 0.

According to the Markov method based on the state-transitions diagram (Fig. 7.1) one can write the following system of differential equations for heat exchanger:

$$\begin{cases} \frac{dp_1^{HE}(t)}{dt} = -\left(4\lambda^{HE} + \lambda^{HEt}\right)p_1^{HE}(t) + \mu^{HE}p_4^{HE}(t) + \mu^{HE}p_5^{HE}(t) \\ \frac{dp_2^{HE}(t)}{dt} = 4\lambda^{HE}p_1^{HE}(t) - \left(3\lambda^{HE} + \lambda^{HEt}\right)p_2^{HE}(t) \\ \frac{dp_3^{HE}(t)}{dt} = 3\lambda^{HE}p_2^{HE}(t) - \left(2\lambda^{HE} + \lambda^{HEt}\right)p_3^{HE}(t) \\ \frac{dp_4^{HE}(t)}{dt} = 2\lambda^{HE}p_3^{HE}(t) - \left(\lambda^{HE} + \lambda^{HEt}\right)p_4^{HE}(t) \\ \frac{dp_5^{HE}(t)}{dt} = \lambda^{HE}p_1^{HE}(t) + \lambda^{HE}p_2^{HE}(t) + \lambda^{HE}p_3^{HE}(t) + \left(\lambda^{HE} + \lambda^{HEt}\right)p_4^{HE}(t) - \mu^{HE}p_5^{HE}(t) \end{cases} \tag{7.15}$$

Initial conditions: $\mathbf{p}^{HE}(0) = \{1, 0, 0, 0, 0\}$.

A numerical solution for states probabilities $p_i^{HE}(t),\ i = 1, \ldots, 5$ can be obtained by solving system (7.15) of differential equations under the given initial conditions by using MATLAB®. Therefore, one obtains the following two sets:

$$\begin{cases} \mathbf{g}^{HE} \in \left\{g_1^{HE}, g_2^{HE}, g_3^{HE}, g_4^{HE}, g_5^{HE}\right\} = \{24, 22, 20, 16, 0\}; \\ \mathbf{p}^{HE}(t) = \{p_1^{HE}(t), p_2^{HE}(t), p_3^{HE}(t), p_4^{HE}(t), p_5^{HE}(t)\}. \end{cases}$$

Having these sets $\mathbf{g}^{HE}$, $\mathbf{p}^{HE}(t)$ one can define L_Z-transform, associated with the heat exchanger's output performance stochastic process $G^{HE}(t)$:

$$L_Z\left\{G^{HE}(t)\right\} = p_1^{HE}(t)z^{24} + p_2^{HE}(t)z^{22} + p_3^{HE}(t)z^{20} + p_4^{HE}(t)z^{16} + p_5^{HE}(t)z^0 \tag{7.16}$$

7.2.2.3 Pump Sub-system

The pump sub-system consists of the pumps, connected in parallel. Each pump supports full cooling capacity of the system. Each pump can be in one of two states: an operational state with capacity 24×10^4 BTU per hour and in state of total failure associated with capacity 0.

Using the state-transitions diagram (Fig. 7.1) in accordance with the Markov method one can write the following system of differential equations for each pump $(i = 1, 2)$:

$$\begin{cases} \dfrac{dp_{i1}^P(t)}{dt} = -\lambda_i p_{i1}^P(t) + \mu_i p_{i2}^P(t), \\ \dfrac{dp_{i2}^P(t)}{dt} = \lambda_i p_{i1}^P(t) - \mu_i p_{i2}^P(t). \end{cases} \quad i = 1, 2 \tag{7.17}$$

Initial conditions are: $p_{i1}^P(0) = 1;\ p_{i2}^P(0) = 0$.

We used MATLAB® for the numerical solution of these systems of differential equations to obtain probabilities $p_{i1}^P(t)$, $p_{i2}^P(t)$, $(i = 1, 2)$

Therefore, for the elements of pump sub-systems, the output performance stochastic processes can be obtained in the following manner:

$$\begin{cases} \mathbf{g}_i^P = \{g_{i1}^P, g_{i2}^P\} = \{24, 0\}, \\ \mathbf{p}_i^P(t) = \{p_{i1}^P(t), p_{i2}^P(t)\}; \end{cases}$$

Sets $\mathbf{g}_i^P$, $\mathbf{p}_i^P(t)(i = 1, 2)$ define L_Z-transforms for each pump as follows:

$$L_Z\left\{g_i^P(t)\right\} = p_{i1}^P(t)z^{g_{i1}^P} + p_{i2}^P(t)z^{g_{i2}^P} = p_{i1}^P(t)z^{24} + p_{i2}^P(t)z^0 \tag{7.18}$$

Using the composition operator Ω_{fpar} for two pumps, connected in parallel, one can obtain the L_Z-transform for output performance stochastic process $G^{Ch}(t)$ of the pump sub-system

$$
\begin{aligned}
L_Z\{G^P(t)\} &= \Omega_{fpar}\left(L_Z\{g_1^P(t)\}, L_Z\{g_2^P(t)\}\right) \\
&= \Omega_{fpar}\left(p_{11}^P(t)z^{24} + p_{12}^P(t)z^0, p_{21}^P(t)z^{24} + p_{22}^P(t)z^0\right) \\
&= P_1^P(t)z^{48} + P_2^P(t)z^{24} + P_3^P(t)z^0.
\end{aligned} \tag{7.19}
$$

where

$$
\begin{aligned}
P_1^P(t) &= p_{11}^P(t)p_{21}^P(t); \\
P_2^P(t) &= p_{11}^P(t)p_{22}^P(t) + p_{21}^P(t)p_{12}^P(t); \\
P_3^P(t) &= p_{12}^P(t)p_{22}^P(t).
\end{aligned}
$$

7.2.3 The L_Z-Transform Application for Calculation of the Reliability Indices of the Water Cooling System

Now the entire system L_Z-transform and the system availability should be found. The system consists of connected in series chiller sub-system, heat exchanger and pump sub-system. Therefore, L_Z-transform for the output performance process of the entire water cooling system $G^S(t)$ can be obtained as follows:

$$
\begin{aligned}
L_Z\{G^S(t)\} &= \Omega_{fser}\left(L_Z\{G^{Ch}(t)\}, L_Z\{g^{HE}(t)\}, L_Z\{G^P(t)\}\right) \\
&= \Omega_{fser}\left(P_1^{Ch}(t)z^{36} + P_2^{Ch}(t)z^{24} + P_3^{Ch}(t)z^{12} + P_4^{Ch}(t)z^0,\right. \\
&\quad p_1^{HE}(t)z^{24} + p_2^{HE}(t)z^{22} + p_3^{HE}(t)z^{20} + p_4^{HE}(t)z^{16} + p_5^{HE}(t)z, \\
&\quad \left.P_1^P(t)z^{48} + P_2^P(t)z^{24} + P_3^P(t)z^0\right)
\end{aligned} \tag{7.20}
$$

After simple algebra, one will obtain the final expression for L_Z-transform of the entire MSS output performance stochastic process $G_S(t)$ in the following form:

$$
\begin{aligned}
L_Z\{G^S(t)\} &= P_1^S(t)z^{g_1^S} + P_2^S(t)z^{g_2^S} + P_3^S(t)z^{g_3^S} + P_4^S(t)z^{g_4^S} + P_5^S(t)z^{g_5^S} + P_6^S(t)z^{g_6^S} \\
&= P_1^S(t)z^{24} + P_2^S(t)z^{22} + P_3^S(t)z^{20} + P_4^S(t)z^{16} + P_5^S(t)z^{12} + P_6^S(t)z^0
\end{aligned} \tag{7.21}
$$

where

$$
\begin{aligned}
P_1^S(t) &= P_1^{Ch}(t)p_1^{HE}(t)P_1^P(t); \\
P_2^S(t) &= P_1^{Ch}(t)p_2^{HE}(t)P_1^P(t); \\
P_3^S(t) &= P_1^{Ch}(t)p_3^{HE}(t)P_1^P(t); \\
P_4^S(t) &= P_1^{Ch}(t)p_4^{HE}(t)P_1^P(t); \\
P_5^S(t) &= P_2^{Ch}(t)\left(1 - p_5^{HE}(t)\right)P_1^P(t);
\end{aligned}
$$

$$P_6^S(t) = \left(\left(P_1^{Ch}(t) + P_2^{Ch}(t)\right)p_5^{HE}(t) + P_3^{Ch}(t)\right)P_1^P(t) + P_2^P(t);$$

Using L_Z-transform (7.21), one obtains MSS instantaneous availability for the MRI water cooling system, working under constant demand. For constant demand level $w = 18 \times 10^4$ BUT/h it will be the following

$$A_{w=18}(t) = \sum_{g_i^S \geq 18} P_i^S(t) = P_1^S(t) + P_2^S(t) + P_3^S(t) \tag{7.22}$$

Using L_Z-transform (7.9) and according to expression (3.29) from the Chap. 3, one obtains mean output MSS performance as following:

$$\begin{aligned}\bar{G}(t) &= \sum_{g_i^S > 0} g_i^S P_i^S(t) \\ &= 24 \cdot P_1^S(t) + 22 \cdot P_2^S(t) + 20 \cdot P_3^S(t) + 16 \cdot P_4^S(t) + 12 \cdot P_5^S(t).\end{aligned} \tag{7.23}$$

Using L_Z-transform (7.21) and according to (3.31) from the Chap. 3, one obtains the expected instantaneous performance deficiency at instant t for constant demand $w = 18 \times 10^4$ BTU/h as following:

$$d_{w=18}^{(E)}(t) = \sum_{i=1}^{6} P_i^S(t)\max\left\{18 - g_i^S, 0\right\} = 2 \cdot P_4^S(t) + 6 \cdot P_5^S(t) + 18 \cdot P_6^S(t). \tag{7.24}$$

The accumulated expected performance deficiency during time interval [0, t] for constant demand $w = 18 \times 10^4$ BTU/h, as a function of time t, can be calculated as follows:

$$\begin{aligned}D_{w=18}(t) &= \int_0^t \sum_{i=1}^{6} P_i^S(x)\max\left\{w - g_i^S, 0\right\}dx \\ &= \int_0^t \left(2 \cdot P_4^S(x) + 6 \cdot P_5^S(x) + 18 \cdot P_6^S(x)\right)dx\end{aligned} \tag{7.25}$$

Reliability indices calculations were performed using the following failure and repair rates.

The failure rate of each chiller is $\lambda^{Ch} = 3.2$ year^{-1}. The repair rate of the each chiller is $\mu^{Ch} = 120$ year^{-1}.

The failure and repair rates corresponding to heat exchanger are: $\lambda^{HE} = 0.1 + 0.33t^2$ year^{-1}, $\lambda^{HEt} = 0.2$ year^{-1}, $\mu^{HE} = 200$ year^{-1}. As one can see, the failure rate of heat exchanger is increasing functions of time, this element possessing the aging property.

The failure rate of each pump is $\lambda^P = 3$ year^{-1}. The repair rate of the pump is $\mu^P = 200$ year^{-1}.

Demand is constant and equal to $w = 18 \times 10^4$ BTU/h.

In Fig. 7.2, one can see a graph of the entire MSS availability $A_{w=18}(t)$ as a function of time t. It can be seen that decreasing availability will down crossing required availability level $A_0 = 0.98$ after 3.36 years. Therefore, replacement of heat exchanger should be performed after 3.36 years. For comparison in Fig. 7.2 is shown another graph, which presents a function $A_{w=18}(t)$ for non-aging case, when heat exchanger failure rate is constant ($\lambda^{HE} = 0.1$ year^{-1}). It can be seen that in this case the system's availability $A_{w=18}(t)$ will be constant, when $t \to \infty$, and equal to 0.996. It will not ever down crossing required availability level 0.98.

In Fig. 7.3, one can see a graph of the mean output MSS performance as a function of time t.

For comparison in Fig. 7.3 is shown another graph, which presents a function $\bar{G}(t)$ for non-aging case, when heat exchanger failure rate is constant ($\lambda^{HE} = 0.1$ year^{-1}). It can be seen, that in this case the system's mean output performance will have constant steady state (when $t \to \infty$) value that is equal 22.3×10^4 BTU/h.

In Fig. 7.4, one can see a graph of the expected instantaneous performance deficiency as a function of time t. For comparison in Fig. 7.4 is shown another graph, which presents a function $d_{w=18}^{(E)}(t)$ for non-aging case, when heat exchanger failure rate is constant ($\lambda^{HE} = 0.1$ year^{-1}). It can be seen, that in this case the system's expected instantaneous performance deficiency after short time is constant function, when for aging system $d_{w=18}^{(E)}(t)$ is increase function.

In Fig. 7.5, one can see a graph of the expected accumulated performance deficiency during time interval [0, t] for constant demand $w = 18 \times 10^4$ BTU/h, as a function of time t.

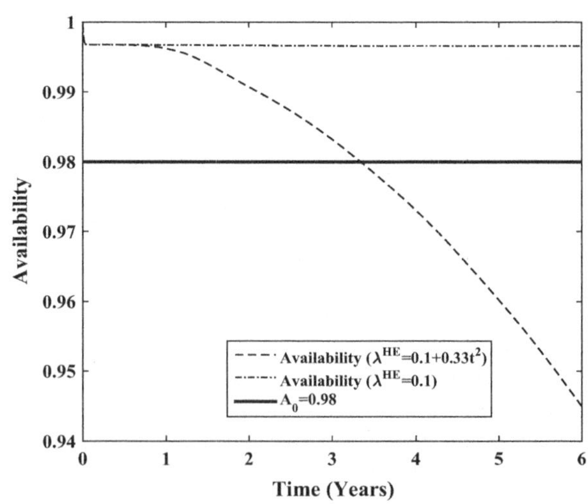

Fig. 7.2 MSS instantaneous availability $A_{w=18}(t)$ for constant demand $w = 18 \times 10^4$ BTU/h

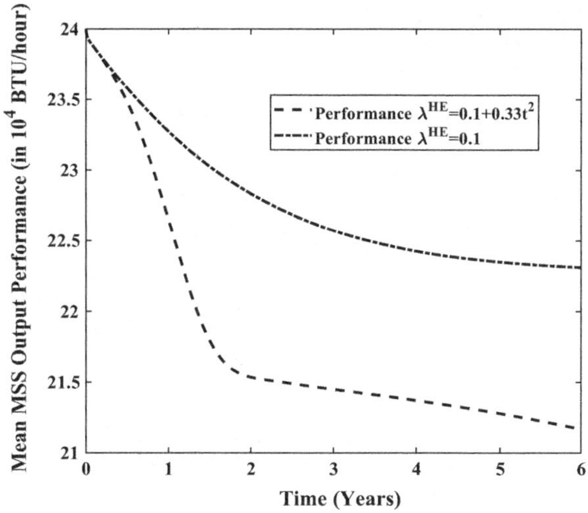

Fig. 7.3 MSS mean instantaneous output performance $\bar{G}(t)$

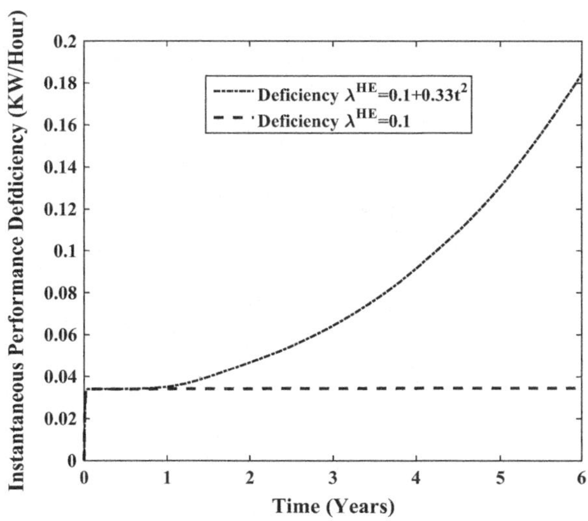

Fig. 7.4 MSS expected instantaneous output performance deficiency $d_{w=18}^{(E)}(t)$ for constant demand $w = 18 \times 10^4$ BTU/h

7.3 Sensitivity Analysis of the Water Cooling System

In Fig. 7.6 are presented graphs of the MSS life time with respect to availability restriction $T_{A_{w=18} \geq 0.98}$ as functions of the chiller failure rate λ^{Ch} and pump failure rate λ^{p}. Note, that when λ^{Ch} is changing, failure rate λ^{p} is keeping its nominal value and when λ^{p} is changing, failure rate λ^{Ch} is keeping its nominal value.

Sensitivity measures with respect to availability restriction were calculated by using expression (7.4): $S_{\lambda^{Ch}}^{A_w} = -0.12$ year^{-2} and $S_{\lambda^{p}}^{A_w} = -0.01$ year^{-2}. It can be

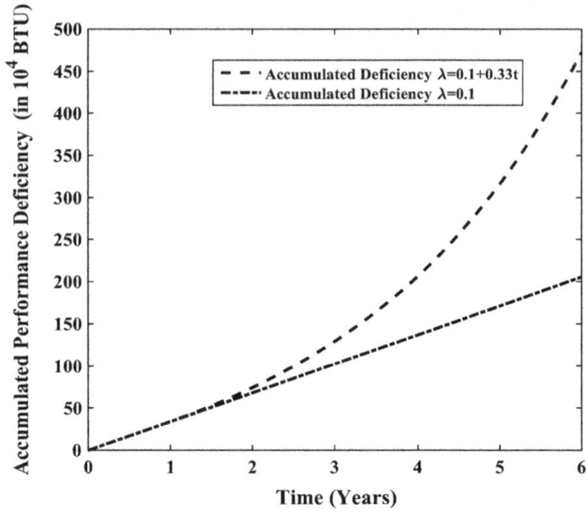

Fig. 7.5 MSS expected accumulated performance deficiency for constant demand $w = 18 \times 10^4$ BTU/h

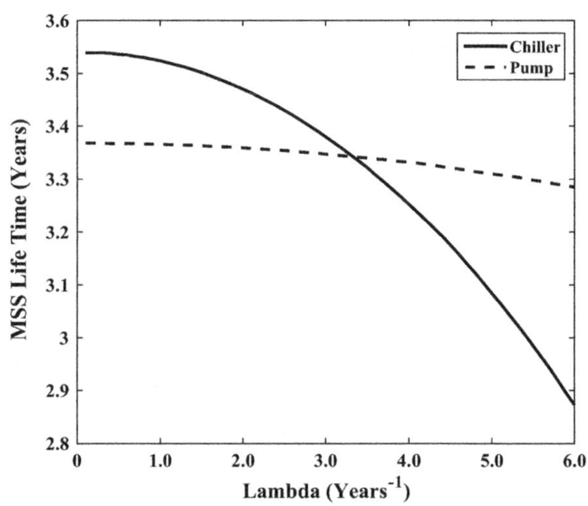

Fig. 7.6 Graphs of the MSS life time with respect to availability restriction $T_{A_{w=18} \geq 0.98}(\lambda^{Ch})$ as functions of the chiller failure rate λ^{Ch} (solid line) and $T_{A_{w=18} \geq 0.98}(\lambda^{P})$ as functions of the pump failure rate λ^{P} (dashed line)

seen that changing of failure rate λ^P has negligible impact to the system life time. At that time, the impact of changing failure rate λ^{Ch} is much greater and it is essential.

In Fig. 7.7 are presented graphs of the MSS life time with respect to mean output performance restriction $T_{\bar{G} \geq 21.5}$ as functions of the chiller failure rate λ^{Ch} and pump failure rate λ^{P}. Note, that when λ^{Ch} is changing, failure rate λ^P is keeping its nominal value and when λ^P is changing, failure rate λ^{Ch} is keeping its nominal value.

Sensitivity measures with respect to mean output performance restriction were calculated by using expression (7.7): $S^{\bar{G}}_{\lambda^{Ch}} = -0.14\ \text{year}^{-2}$ and $S^{\bar{G}}_{\lambda^{P}} = -0.04\ \text{year}^{-2}$. It can be seen that, as in the previous case, the impact of changing failure rate λ^{Ch} is much greater than impact of changing failure rate λ^{P}.

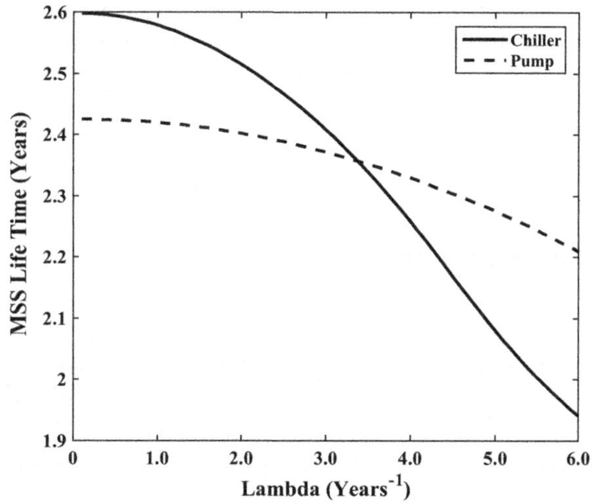

Fig. 7.7 Graphs of the MSS life time with respect to mean output performance restriction $T_{\bar{G}\geq 21.5}(\lambda^{Ch})$ as functions of the chiller failure rate λ^{Ch} (solid line) and $T_{\bar{G}\geq 21.5}(\lambda^{P})$ as functions of the pump failure rate λ^{P} (dashed line)

In Fig. 7.8 are presented graphs of the MSS life time with respect to accumulated performance deficiency during time interval [0, t] for constant demand $w = 18 \times 10^4$ BTU/h $T_{D_{w=18}\leq 150}$ as functions of the chiller failure rate λ^{Ch} and pump failure rate λ^{P}. Note, that when λ^{Ch} is changing, failure rate λ^{p} is keeping its nominal value and when λ^{p} is changing, failure rate λ^{Ch} is keeping its nominal value.

Sensitivity measures with respect to accumulated performance deficiency were calculated by using expression (7.11, 7.12): $S^{D_w}_{\lambda^{Ch}} = -0.34$ year^{-2} and $S^{D_w}_{\lambda^{P}} = -0.12$ year^{-2}.

It can be seen that, as in previous cases, the impact of changing failure rate λ^{Ch} is much greater than the impact of changing failure rate λ^{p}.

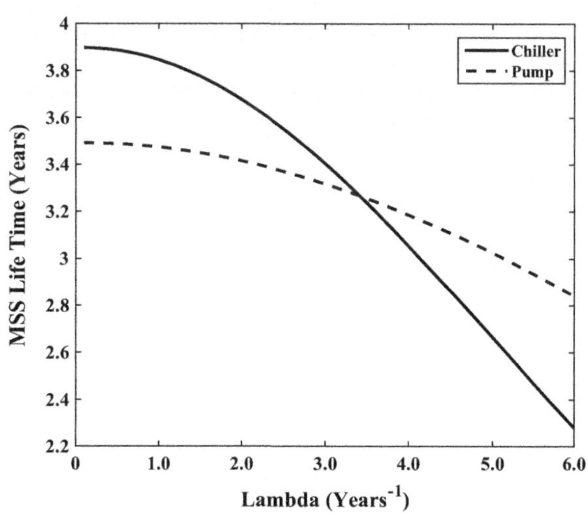

Fig. 7.8 Graphs of the MSS life time with respect to accumulated performance deficiency during time interval [0, t] for constant demand w = 18 × 10^4 BTU/h $T_{D_{w=18}\leq 150}(\lambda^{Ch})$ as functions of the chiller failure rate λ^{Ch} (solid line) and $T_{D_{w=18}\leq 150}(\lambda^{P})$ as functions of the pump failure rate λ^{P} (dashed line)

In practice, for reliability engineer, it means that he should be very accurate, when orders chillers' equipment from a supplier. It would be recommended to organize for chiller a special reliability field demonstration test, in order to justify chiller's nominal failure rate $\lambda^{Ch} = 3.2$ year^{-1} and prevent a delivery of less reliable chillers (with $\lambda^{Ch} > 3.2$ year^{-1}).

7.4 Summary

In the chapter, the method for analysis of impact of changing failure or repair rates in different elements in aging MSS is considered.

The chapter introduces sensitivity measures that are useful for investigation of an aging MSS and presents the method for their evaluation, based on using L_Z-transform. As sensitivity measures were proposed derivations of MSS life time with respect to restriction of availability, mean output performance and accumulated performance deficiency.

References

1. ASHRAE Handbook (2014) Refrigeration. American Society of Heating Refrigerating and Air-Conditioning Engineers, Atlanta
2. Lisnianski A (2012) *Lz*-transform for a discrete-state continuous-time Markov process and its application to multi-state system reliability. In: Lisnianski A, Frenkel I (eds) Recent advances in system reliability. Signatures, multi-state systems and statistical inference. Springer, London, pp 79–95
3. Lisnianski A, Frenkel I, Ding Y (2010) Multi-state system reliability analysis and optimization for engineers and industrial managers. Springer, London
4. Lisnianski A, Frenkel I, Khvatskin L (2017) On sensitivity analysis of aging multi-state system using the Lz-transform method. Reliab Eng Syst Saf 166:99–108
5. Saltelli A, Chan K, Scott EM (2009) Sensitivity analysis. Wiley, Chichester
6. Trivedi K (2002) Probability and statistics with reliability, queuing and computer science applications. Wiley, New York

Chapter 8
Birnbaum Importance Assessment for Aging Multi-state Water Cooling System

Methods for evaluating the influence of the reliability of specific elements on the reliability or availability of the entire system provide useful information about the importance of these elements. Importance evaluation is an essential point in tracing bottlenecks in systems and in the identification of the most important elements. It is a useful tool to help the analyst find weaknesses in design and to suggest modifications for system upgrades. Importance was firstly introduced by Birnbaum [1] and later became known as B-reliability importance [4]. This index characterizes the rate at which system reliability changes with respect to changes in the reliability of a given element. An improvement in reliability of an element with the highest importance causes the greatest increase in system reliability. In [4] one can find most complete presentation of the state of the art in the field.

In this chapter the L_Z-transform is applied to the dynamic importance analysis of an aging MSS composed of multi-state and binary elements with different performances under minimal repair [2, 7]. This method also takes demand into account. It was primarily introduced in [8], where one can find an extended reference list corresponding to the issue. Steady-state importance analysis for binary state elements in MSS was presented in [5].

The method suggested for dynamic importance analysis essentially simplifies a computational burden either in comparison with the straightforward Markov method or with simulation.

8.1 Dynamic B-Availability Importance Assessment by Using the L_Z-Transform Method

As we know from Chap. 3, if the L_Z-transform of the entire output stochastic process $G(t) \in \{g_1, \ldots, g_K\}$ of the MSS is known, then important system reliability measures can be easily obtained.

A. Lisnianski et al., *Modern Dynamic Reliability Analysis for Multi-state Systems*, Springer Series in Reliability Engineering,
https://doi.org/10.1007/978-3-030-52488-3_8

$$L_Z\{G(t)\} = \sum_{k=1}^{K} p_k(t) z^{g_k} \tag{8.1}$$

The system availability for the demand level w at instant $t \geq 0$ is defined as follows:

$$A_w(t) = \sum_{g_k \geq w} p_k(t). \tag{8.2}$$

By differentiating (8.2) one can obtain dynamic B-availability importance for each binary state element j

$$I_{A_w}^{(j)}(t) = \frac{\partial A_w(t)}{\partial p_j(t)}, \tag{8.3}$$

where $p_j(t)$ is the probability that the j-th binary-state element will be in the state of perfect functioning with performance g_{j1} at time instant $t \geq 0$.

8.2 B-Availability Importance Assessment for the Water Cooling System of Magnetic Resonance Inspection Equipment

8.2.1 System Description and Corresponding Reliability Data

We consider a magnetic resonance inspection (MRI) water cooling system that is used in hospitals [3]. For MRI scanners to operate at the highest levels of efficiency the magnet inside the scanner must be kept as cool as possible. The water cooling system analyzed in this Chapter is responsible for removing heat from the MRI scanner. The system is similar to the water cooling system that was presented in the previous Chapter, but there is the essential difference. In the presented Chapter is considered the system where heat exchanger subsystem consists of three different aging heat exchangers. Such configuration is also widely used in practice (In the previous Chapter was considered a water cooling system with one complex heat exchanger, which is presented as multi-state element with 5 states).

The structure of the water cooling system and the state-transition diagram for each of its elements are presented in Fig. 8.1.

As one can see in the system only two elements, Chiller Controller and Pump Controller are multi-state elements. All other elements are binary-state elements.

The nominal performance of the entire system is $24 \cdot 10^4$ BTU/h (BTU-British Thermal Unit).

The chillers sub-system consists of three chillers, connected in parallel and their performance can be allowed in only one of two principal states: full capacity or zero

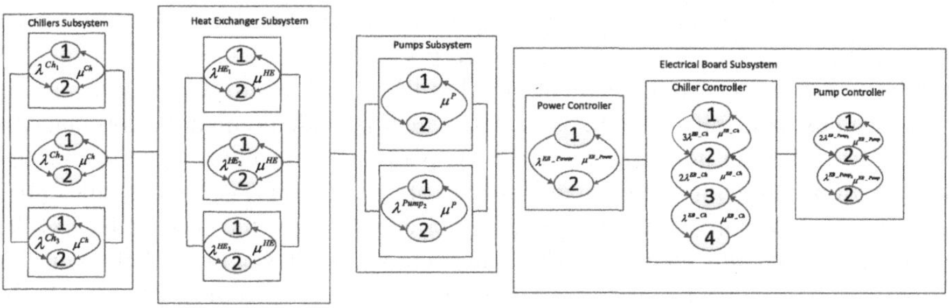

Fig. 8.1 Structure and state-transition diagram of the water cooling system

capacity. According to the equipment type, the cooling capacity of each chiller is $12 \cdot 10^4$ BTU/h; so two simultaneously working chillers can cover a maximal cooling load, which is needed for effective MRI functioning.

The failure rate of each chiller is $\lambda_i^{Ch} = 3$ year^{-1}. The repair rate of the each chiller is $\mu^{Ch} = 365$ year^{-1}.

The heat exchanger subsystem consists of three heat exchangers, connected in parallel. The cooling capacities of heat exchangers is as follows: the first one has $16 \cdot 10^4$ BTU/h, the second and the third heat exchangers have the same cooling capacity—$4 \cdot 10^4$ BTU/h. So, heat exchanger subsystem has 4 states of performance degradation: a fully operational state with a capacity of $24 \cdot 10^4$ BTU/h (all heat exchangers are fully operated), states of partial failure corresponding to capacities of $20 \cdot 10^4$ and $16 \cdot 10^4$ BTU per hour, and a total failure corresponding to a capacity less then $16 \cdot 10^4$ BTU per hour and equal to 0. All heat exchangers have the same failure and repair rates and they are: $\lambda_i^{HE} = 0.1 + t^2$ year^{-1}, $i = 1, 2, 3$ and $\mu^{HE} = 200$ year^{-1}. As one can see, the heat exchangers are aging elements because they failure rates are an increasing functions.

Cold-water circulation is realized by two pumps. Each pump supports the nominal cooling capacity of the entire system. In case of failure of the first pump, the second automatically begins its operation. Only after the breakdown of both pumps does the system fail completely. Each pump can be in one of two states: a fully operational state, which provides the whole (entire) MSS with a nominal performance of $24 \cdot 10^4$ BTU per hour, and a state of total failure equal to 0. The failure rate of the first pump is $\lambda^{Pump_1} = 3$ year^{-1}, while the failure rate of the second pump is $\lambda^{Pump_2} = 6$ year^{-1}. The repair rate of the pumps is $\mu^{Pump} = 365$ year^{-1}.

The principal purpose of the electrical board is power supply and control of the different sub-systems of the cooling system. It is possible to present the electrical board as a system with three parts connected in series:

- The first part of the electrical board is the power controller, which can be in one of two states: a fully operational state, which provides a performance level $24 \cdot 10^4$ BTU per hour to the entire system, and a state of total (complete) failure, with the corresponding performance of the entire system equal to zero (0). The failure and repair rates of the power controller in the electrical board are $\lambda^{EB\_Power} = 1$ year^{-1} and $\mu^{EB\_Power} = 1000$ year^{-1}, respectively.

- The second part of the electrical board is the chiller controller, which may be in 4 states: two fully operational states, which provide nominal performance of $36 \cdot 10^4$ BTU per hour and $24 \cdot 10^4$ BTU per hour to the entire system, a partial operation state with nominal performance of $12 \cdot 10^4$ BTU per hour to the entire system, and a state of complete failure, corresponding to the performance of the entire system being equal to 0. The failure and repair rates of the chiller controller are $\lambda^{EB\_Ch} = 2$ year^{-1} and $\mu^{EB\_Ch} = 1000$ year^{-1}, respectively.
- The third part of the electrical board is the pump controller, which has 3 states: two fully operational states, which provides a performance of $48 \cdot 10^4$ BTU per hour and $24 \cdot 10^4$ BTU per hour for the entire system, and a state of total failure corresponding to the performance of the entire system being equal to 0. The failure and repair rates of pump controller sub-system are $\lambda^{EB\_Pump} = 2$ year^{-1} and $\mu^{EB\_Pump} = 1000$ year^{-1}.

The problem is to find B-availability importance for chillers and heat exchangers as functions of time t for different demand levels $w_1 = 24 \cdot 10^4$ BTU/h, $w_2 = 20 \cdot 10^4$ BTU/h, $w_3 = 16 \cdot 10^4$ BTU/h. From now, we shall omit the 10^4 multiplier for cooling capacity and write, for example, 24, instead of $24 \cdot 10^4$ BTU/h.

So, accordingly with expression (8.3) we should find the B-availability importance [4] of the first chiller:

$$I_{A_{w=24}}^{Ch,1}(t) = \frac{\partial A_{w=24}(t)}{\partial p_{11}^{Ch}(t)};\ I_{A_{w=20}}^{Ch,1}(t) = \frac{\partial A_{w=20}(t)}{\partial p_{11}^{Ch}(t)};\ I_{A_{w=16}}^{Ch,1}(t) = \frac{\partial A_{w=16}(t)}{\partial p_{11}^{Ch}(t)}, \tag{8.4}$$

where $p_{11}^{Ch}(t)$ is the probability that first chiller will be in given state 1 with performance $12 \cdot 10^4$ BTU/h and demand of $24 \cdot 10^4$ BTU/h, $20 \cdot 10^4$ BTU/h and $16 \cdot 10^4$ BTU/h respectively, at time instant $t \geq 0$.

Similarity, the B-availability importance of the heat exchangers will be presented as follows:

$$I_{A_{w=24}}^{HE,i}(t) = \frac{\partial A_{w=24}(t)}{\partial p_{i1}^{HE}(t)};\ I_{A_{w=20}}^{HE,i}(t) = \frac{\partial A_{w=20}(t)}{\partial p_{i1}^{HE}(t)};\ I_{A_{w=16}}^{HE,i}(t) = \frac{\partial A_{w=16}(t)}{\partial p_{i1}^{HE}(t)},\ i = 1, 2, 3, \tag{8.5}$$

where p_{i1}^{HE} is the probability that i-th heat exchanger will be in given state 1 (with full performance) at time instant $t \geq 0$.

8.2.2 Computation L_Z-Transforms for System's Elements and Sub-systems

According to the suggested method, the first step of the solution is finding L_Z-transforms for all subsystems [6].

8.2.2.1 L_Z-Transform for the Chillers Sub-system

Using the state-transitions diagram (Fig. 8.1) in accordance with the Markov method, we build the following system of differential equations [8] for each chiller ($i = 1, 2, 3$):

$$\begin{cases} \frac{dp_{i1}^{Ch}(t)}{dt} = -\lambda^{Ch} p_{i1}^{Ch}(t) + \mu^{Ch} p_{i2}^{Ch}(t), \\ \frac{dp_{i2}^{Ch}(t)}{dt} = \lambda^{Ch} p_{i1}^{Ch}(t) - \mu^{Ch} p_{i2}^{Ch}(t). \end{cases} \tag{8.6}$$

with the following initial conditions:

$$p_{i1}^{Ch}(0) = 1; \ p_{i2}^{Ch}(0) = 0. \tag{8.7}$$

An analytical solution of the system of differential equations (8.6) for probabilities $p_{i1}^{Ch}(t)$ and $p_{i2}^{Ch}(t)$ ($i = 1, 2, 3$) under initial conditions (8.7) is as follows:

$$\begin{cases} p_{i1}^{Ch} = {}^{365}/_{368} + {}^{3}/_{368}e^{-368t}; \\ p_{i2}^{Ch} = {}^{3}/_{368} - {}^{3}/_{368}e^{-368t}. \end{cases} \tag{8.8}$$

Therefore, for each chiller i, $i = 1, 2, 3$ we have the set of states for its output performance stochastic processes $\mathbf{g}_i^{Ch} = \{g_{i1}^{Ch}, g_{i2}^{Ch}\} = \{12, 0\}$, and the probabilities $\mathbf{p}_{i1}^{Ch}(t) = \{p_1^H(t), p_2^H(t)\}$ of these states as functions of time t.

Based on these findings one can define the L_Z-transform for each chiller's output performance stochastic process $G_i^{Ch}(t)$, as follows:

$$L_Z\{G_i^{Ch}(t)\} = p_{i1}^{Ch}(t)z^{g_{i1}^{Ch}} + p_{i2}^{Ch}(t)z^{g_{i2}^{Ch}} = p_{i1}^{Ch}(t)z^{12} + p_{i2}^{Ch}(t)z^{0}, \ i = 1, 2, 3. \tag{8.9}$$

Using the composition operator Ω_{fpar} [7, 9] for 3 chillers, connected in parallel, one can obtain the L_Z-transform $L_Z\{G^{Ch}(t)\}$ for the chiller sub-system, as a multiplication of individual L_Z-transforms of all 3 chillers:

$$L_Z\{G^{Ch}(t)\} = \Omega_{fpar}\{L_Z\{G_1^{Ch}(t)\}, L_Z\{G_2^{Ch}(t)\}, L_Z\{G_3^{Ch}(t)\}\}$$

$$
\begin{aligned}
&= \Omega_{fpar}\left\{ \begin{array}{l} p_{11}^{Ch}(t)z^{g_{11}^{Ch}} + p_{12}^{Ch}(t)z^{g_{12}^{Ch}},\ p_{21}^{Ch}(t)z^{g_{21}^{Ch}} \\ +p_{22}^{Ch}(t)z^{g_{22}^{Ch}},\ p_{31}^{Ch}(t)z^{g_{31}^{Ch}} + p_{32}^{Ch}(t)z^{g_{32}^{Ch}} \end{array} \right\} \\
&= \Omega_{fpar}\left\{ \begin{array}{l} p_{11}^{Ch}(t)z^{12} + p_{12}^{Ch}(t)z^{0},\ p_{21}^{Ch}(t)z^{12} \\ +p_{22}^{Ch}(t)z^{0},\ p_{31}^{Ch}(t)z^{12} + p_{32}^{Ch}(t)z^{0} \end{array} \right\} \\
&= p_{11}^{Ch}(t)p_{21}^{Ch}p_{31}^{Ch}(t)z^{36} \\
&\quad \left[\begin{array}{r} p_{11}^{Ch}(t)p_{21}^{Ch}(t)p_{32}^{Ch}(t) + p_{11}^{Ch}(t)p_{22}^{Ch}(t)p_{31}^{Ch}(t) \\ +p_{12}^{Ch}(t)p_{21}^{Ch}(t)p_{31}^{Ch}(t) \end{array} \right] z^{24} \\
&\quad \left[\begin{array}{r} p_{11}^{Ch}(t)p_{22}^{Ch}(t)p_{32}^{Ch}(t) + p_{12}^{Ch}(t)p_{21}^{Ch}(t)p_{32}^{Ch}(t) \\ +p_{12}^{Ch}(t)p_{22}^{Ch}(t)p_{31}^{Ch}(t) \end{array} \right] z^{12} \\
&\quad + p_{12}^{Ch}(t)p_{22}^{Ch}(t)p_{32}^{Ch}(t)z^{0}
\end{aligned}
\tag{8.10}
$$

According to technical conditions, we need only two chillers from three existing, so we can combine states $36 \cdot 10^4$ BTU/h and $24 \cdot 10^4$ BTU/h into common state $24 \cdot 10^4$ BTU/h and write expression (8.10) in the following form

$$
L_Z\{G^{Ch}(t)\} = P_1^{Ch}(t)z^{24} + P_2^{Ch}(t)z^{12} + P_3^{Ch}(t)z^{0}, \tag{8.11}
$$

where

$$
\begin{aligned}
P_1^{Ch}(t) &= p_{11}^{Ch}(t)p_{21}^{Ch}(t)p_{31}^{Ch}(t) + p_{11}^{Ch}(t)p_{21}^{Ch}(t)p_{32}^{Ch}(t) + p_{11}^{Ch}(t)p_{22}^{Ch}(t)p_{31}^{Ch}(t) \\
&\quad + p_{12}^{Ch}(t)p_{21}^{Ch}(t)p_{31}^{Ch}(t) \\
&= p_{21}^{Ch}(t)p_{31}^{Ch}(t) + p_{11}^{Ch}(t)p_{21}^{Ch}(t)p_{32}^{Ch}(t) + p_{11}^{Ch}(t)p_{22}^{Ch}(t)p_{31}^{Ch}(t) \\
P_2^{Ch}(t) &= p_{11}^{Ch}(t)p_{22}^{Ch}(t)p_{32}^{Ch}(t) + p_{12}^{Ch}(t)p_{21}^{Ch}(t)p_{32}^{Ch}(t) \\
&\quad + p_{12}^{Ch}(t)p_{22}^{Ch}(t)p_{31}^{Ch}(t); \\
P_3^{Ch}(t) &= p_{12}^{Ch}(t)p_{22}^{Ch}(t)p_{32}^{Ch}(t).
\end{aligned}
$$

8.2.2.2 L_Z-Transform for the Heat Exchanger Sub-system

Using the state-transitions diagram (Fig. 8.1) in accordance with the Markov method, we build the following system of differential equations [9] for each heat exchanger ($i = 1, 2, 3$):

$$
\begin{cases}
\dfrac{dp_{i1}^{HE}(t)}{dt} = -\lambda_i^{HE} p_{i1}^{HE}(t) + \mu^{HE} p_{i2}^{HE}(t), \\
\dfrac{dp_{i2}^{HE}(t)}{dt} = \lambda_i^{HE} p_{i1}^{HE}(t) - \mu^{HE} p_{i2}^{HE}(t).
\end{cases}
\tag{8.12}
$$

with the following initial conditions:

$$p_{i1}^{HE}(0) = 1; \; p_{i2}^{HE}(0) = 0. \tag{8.13}$$

An explicit analytical solution for probabilities $p_{i1}^{HE}(t)$ and $p_{i2}^{HE}(t)$ $(i = 1, 2, 3)$ could not be found. Numerical solution was obtained by using MATLAB® for each of these 3 systems of differential equations.

Therefore, for each heat exchanger i, $i = 1, 2, 3$ we have the set of states for its output performance stochastic processes:

$$\begin{aligned} \mathbf{g}_1^{HE} &= \{g_{11}^{HE}, g_{12}^{HE}\} = \{16, 0\}; \\ \mathbf{g}_2^{HE} &= \{g_{21}^{HE}, g_{22}^{HE}\} = \{4, 0\}; \\ \mathbf{g}_3^{HE} &= \{g_{31}^{HE}, g_{32}^{HE}\} = \{4, 0\} \end{aligned} \tag{8.14}$$

and the probabilities $p_{i1}^{HE}(t)$, $p_{i2}^{HE}(t)$ of these states as functions of time t. The L_Z-transforms for each heat exchanger's output performance stochastic process $G_i^{HE}(t)$, can be presented as follows:

$$\begin{aligned} L_Z\{G_1^{HE}(t)\} &= p_{11}^{HE}(t)z^{g_{i1}^{HE}} + p_{12}^{HE}(t)z^{g_{i2}^{HE}} = p_{11}^{HE}(t)z^{16} + p_{12}^{HE}(t)z^{0}, \\ L_Z\{G_2^{HE}(t)\} &= p_{21}^{HE}(t)z^{g_{21}^{HE}} + p_{22}^{HE}(t)z^{g_{22}^{HE}} = p_{21}^{HE}(t)z^{4} + p_{22}^{HE}(t)z^{0}, \\ L_Z\{G_3^{HE}(t)\} &= p_{31}^{HE}(t)z^{g_{31}^{HE}} + p_{32}^{HE}(t)z^{g_{32}^{HE}} = p_{31}^{HE}(t)z^{4} + p_{32}^{HE}(t)z^{0}. \end{aligned} \tag{8.15}$$

Using the composition operator Ω_{fpar} [7, 10] for three heat exchangers, connected in parallel, one can obtain the L_Z-transform $L_Z\{G^{HE}(t)\}$ for the heat exchanger subsystem, as a multiplication of individual L_Z-transforms of all three heat exchangers:

$$\begin{aligned} L_Z\{G^{HE}(t)\} &= \Omega_{fpar}\{L_Z\{G_1^{HE}(t)\}, L_Z\{G_2^{HE}(t)\}, L_Z\{G_3^{HE}(t)\}\} \\ &= \Omega_{fpar}\left\{ \begin{array}{l} p_{11}^{HE}(t)z^{16} + p_{12}^{HE}(t)z^{0},\; p_{21}^{HE}(t)z^{4} \\ +p_{22}^{HE}(t)z^{0},\; p_{31}^{HE}(t)z^{4} + p_{32}^{HE}(t)z^{0} \end{array} \right\} \\ &= p_{11}^{HE}(t)p_{21}^{HE}(t)p_{31}^{HE}(t)z^{24} \\ &\quad + p_{11}^{HE}(t)\left(p_{22}^{HE}(t)p_{31}^{HE}(t) + p_{21}^{HE}(t)p_{32}^{HE}(t)\right)z^{20} \\ &\quad + p_{11}^{HE}(t)p_{22}^{HE}(t)p_{32}^{HE}(t)z^{16} \\ &\quad + p_{12}^{HE}z^{0}. \end{aligned} \tag{8.16}$$

We shall write expression (8.16) in the following form

$$L_Z\{G^{HE}(t)\} == P_1^{HE}(t)z^{24} + P_2^{HE}(t)z^{20} + P_3^{HE}(t)z^{16} + P_4^{HE}(t)z^{0}, \tag{8.17}$$

where

$$P_1^{HE}(t) = p_{11}^{HE}(t)p_{21}^{HE}(t)p_{31}^{HE}(t);$$
$$P_2^{HE}(t) = p_{11}^{HE}(t)\left(p_{22}^{HE}(t)p_{31}^{HE}(t) + p_{21}^{HE}(t)p_{32}^{HE}(t)\right);$$
$$P_3^{HE}(t) = p_{11}^{HE}(t)p_{22}^{HE}(t)p_{32}^{HE}(t);$$
$$P_4^{HE}(t) = p_{12}^{HE}(t).$$

8.2.2.3 L_Z-Transform for Pumps Sub-system

Each pump can support the full cooling capacity of the system. In case of failure of one pump the second automatically turns on. Therefore, complete system failure can occur only after the failure of both pumps.

According to the Markov method and the state-transitions diagram (Fig. 8.1) we obtain the following system of differential equations [8] for each pump ($i = 1, 2$):

$$\begin{cases} \dfrac{dp_{i1}^{Pump}(t)}{dt} = -\lambda^{Pump} p_{i1}^{Pump}(t) + \mu^{Pump} p_{i2}^{Pump}(t), \\ \dfrac{dp_{i2}^{Pump}(t)}{dt} = \lambda^{Pump} p_{i1}^{Pump}(t) - \mu^{Pump} p_{i2}^{Pump}(t). \end{cases} \tag{8.18}$$

Initial conditions are:

$$p_{i1}^{Pump}(0) = 1;\ p_{i2}^{Pump}(0) = 0. \tag{8.19}$$

An analytical solution of the system of differential equations (8.18) under initial conditions (8.19) for probabilities $p_{i1}^{Pump}(t)$ and $p_{i2}^{Pump}(t)$ ($i = 1, 2$) is as follows:

$$\begin{cases} p_{i1}^{Pump} = {}^{365}/_{368} + {}^{3}/_{368}e^{-368t}; \\ p_{i2}^{Pump} = {}^{3}/_{368} - {}^{3}/_{368}e^{-368t}. \end{cases} \tag{8.20}$$

Therefore, for each pump i, $i = 1, 2$, we obtain the following output performance stochastic processes $G_i^{Pump}(t)$, which have two possible states:

$$\begin{aligned} \mathbf{g}_i^{Pump} &= \left\{g_{i1}^{Pump}, g_{i2}^{Pump}\right\} = \{24, 0\} \\ \mathbf{p}_i^{Pump} &= \left\{p_{i1}^{Pump}(t), p_{i2}^{Pump}(t)\right\} \end{aligned} \tag{8.21}$$

The L_Z-transform for each pump, associated with the output performance stochastic process $G_i^{Pump}(t)$ for each pump can be presented in the following form:

$$\begin{aligned} L_Z\left\{G_i^{pump}(t)\right\} = p_{i1}^{Pump}(t)z^{g_{i1}^{Pump}} + p_{i2}^{Pump}(t)z^{g_{i2}^{Pump}} = p_{i1}^{Pump}(t)z^{24} \\ + p_{i2}^{Pump}(t)z^{0},\ i = 1, 2. \end{aligned} \tag{8.22}$$

Using Ushakov's UGO Ω_{fpar} [7, 9] for 2 pumps connected in parallel, one obtains the L_Z-transform for the entire pump sub-system $L_Z\{G^{Pump}(t)\}$, as a multiplication of the individual L_Z-transforms of 2 pumps:

$$\begin{aligned} L_Z\{G^{Pump}(t)\} &= \Omega_{fpar}\left\{L_Z\left\{G_1^{Pump}(t)\right\}, L_Z\left\{G_2^{Pump}(t)\right\}\right\} \\ &= \Omega_{fpar}\left\{\begin{array}{l} p_{11}^{Pump}(t)z^{g_{11}^{Pump}} + p_{12}^{Pump}(t)z^{g_{12}^{Pump}}, \\ p_{21}^{Pump}(t)z^{g_{21}^{Pump}} + p_{22}^{Pump}(t)z^{g_{22}^{Pump}} \end{array}\right\} \\ &= \Omega_{fpar}\left\{p_{11}^{Pump}(t)z^{24} + p_{12}^{Pump}(t)z^{0}, p_{21}^{Pump}(t)z^{24} + p_{22}^{Pump}(t)z^{0}\right\}. \end{aligned} \tag{8.23}$$

According to technical conditions, each pump can support the full cooling capacity of the system, so we can combine states $48 \cdot 10^4$ BTU/h and $24 \cdot 10^4$ BTU/h into common state $24 \cdot 10^4$ BTU/h and after multiplication, one obtains the resulting L_Z-transform for the complete pump sub-system:

$$L_Z\{G^{Pump}(t)\} = P_1^{Pump}(t)z^{24} + P_2^{Pump}(t)z^{0}, \tag{8.24}$$

where

$$\begin{aligned} P_1^{Pump}(t) &= p_{11}^{Pump}(t)p_{21}^{Pump}(t) + p_{11}^{Pump}(t)p_{22}^{Pump}(t) + p_{21}^{Pump}(t)p_{12}^{Pump}(t) \\ &= p_{11}^{Pump}(t) + p_{21}^{Pump}(t)p_{12}^{Pump}(t); \\ P_2^{Pump}(t) &= p_{12}^{Pump}(t)p_{22}^{Pump}(t). \end{aligned}$$

8.2.2.4 L_Z-Transform for Electrical Board Sub-system

According to the state-transition diagram of the electrical board subsystem, there are two states for the stochastic process $G^{EB\_Power}(t)$ performance of the electrical power controller:

$$G^{EB\_Power}(t) \in \mathbf{g}^{EB\_Power} = \left\{g_1^{EB\_Power}, g_2^{EB\_Power}\right\} = \{24, 0\}.$$

By using the Markov method one obtains the following system of differential equations [9] for probabilities of these states:

$$\left\{\begin{array}{l} \dfrac{dp_1^{EB\_Power}(t)}{dt} = -\lambda^{EB\_Power} p_1^{EB\_Power}(t) + \mu^{EB\_Power} p_2^{EB\_Power}(t), \\ \dfrac{dp_2^{EB\_Power}(t)}{dt} = \lambda^{EB\_Power} p_1^{EB\_Power}(t) - \mu^{EB\_Power} p_2^{EB\_Power}(t). \end{array}\right. \tag{8.25}$$

Initial conditions are:

$$p_1^{EB\_Power}(0) = 1;\ p_2^{EB\_Power}(0) = 0. \tag{8.26}$$

An analytical solution of the system of differential equations (8.25) under initial conditions (8.26) for probabilities $\mathbf{p}^{EB\_Power}(t) = \left\{p_1^{EB\_Power}(t),\ p_2^{EB\_Power}(t)\right\}$ is as follows:

$$\begin{cases} p_1^{EB\_Power} = {}^{1000}\!/_{1001} + {}^{1}\!/_{1001}e^{-1001t}; \\ p_2^{EB\_Power} = {}^{1}\!/_{1001} - {}^{1}\!/_{1001}e^{-1001t}. \end{cases} \tag{8.27}$$

Having the sets $\mathbf{g}^{EB\_Power}$, $\mathbf{p}^{EB\_Power}(t)$, one can define the L_Z-transform that is associated with the output performance stochastic process of the power controller:

$$\begin{aligned} L_Z\{G^{EB\_Power}(t)\} &= p_1^{EB\_Power}(t)z^{g_1^{EB\_Power}} + p_2^{EB\_Power}(t)z^{g_2^{EB\_Power}} \\ &= p_1^{EB\_Power}(t)z^{24} + p_2^{EB\_Power}(t)z^{0}. \end{aligned} \tag{8.28}$$

According to the system description one can obtain the output performance stochastic process for the power controller of the chiller.

$$\begin{aligned} \mathbf{g}^{EB\_Ch} &= \left\{g_1^{EB\_Ch}, g_2^{EB\_Ch}, g_3^{EB\_Ch}, g_4^{EB\_Ch}\right\} = \{36,\ 24,\ 12,\ 0\}. \\ \mathbf{p}^{EB\_Ch}(t) &= \left\{p_1^{EB\_Ch}(t),\ p_2^{EB\_Ch}(t),\ p_3^{EB\_Ch}(t),\ p_4^{EB\_Ch}(t)\right\} \end{aligned} \tag{8.29}$$

Using a similar procedure we build the following system of differential equations [8] for the power controller of the chiller:

$$\begin{cases} \frac{dp_1^{EB\_Ch}(t)}{dt} = -3\lambda^{EB\_Ch}p_1^{EB\_Ch}(t) + \mu^{EB\_Ch}p_2^{EB\_Ch}(t), \\ \frac{dp_2^{EB\_Ch}(t)}{dt} = 3\lambda^{EB\_Ch}p_1^{EB\_Ch}(t) - \left(2\lambda^{EB\_Ch} + \mu^{EB\_Ch}\right)p_2^{EB\_Ch}(t) + \mu^{EB\_Ch}p_3^{EB\_Ch}(t), \\ \frac{dp_3^{EB\_Ch}(t)}{dt} = 2\lambda^{EB\_Ch}p_2^{EB\_Ch}(t) - \left(\lambda^{EB\_Ch} + \mu^{EB\_Ch}\right)p_3^{EB\_Ch}(t) + \mu^{EB\_Ch}p_4^{EB\_Ch}(t), \\ \frac{dp_4^{EB\_Ch}(t)}{dt} = \lambda^{EB\_Ch}p_3^{EB\_Ch}(t) - \mu^{EB\_Ch}p_4^{EB\_Ch}(t). \end{cases} \tag{8.30}$$

Initial conditions are:

$$p_1^{EB\_Ch}(0) = 1;\ p_2^{EB\_Ch}(0) = p_3^{EB\_Ch}(0) = p_4^{EB\_Ch}(0) = 0. \tag{8.31}$$

Explicit analytical solutions for the chiller controller may be found, but have very complicated difficult for usage form. A numerical solution for states' probabilities was obtained by using MATLAB for solving system (8.30) under initial conditions (8.31).

Having the sets $\mathbf{g}^{EB\_Ch}$, $\mathbf{p}^{EB\_Ch}(t)$, one can define for the chiller controller the L_Z-transform that is associated with the output performance stochastic process for the power controller:

$$L_Z\left\{G^{EB\_Ch}(t)\right\} = p_1^{EB\_Ch}(t)z^{g_1^{EB\_Ch}} + p_2^{EB\_Ch}(t)z^{g_2^{EB\_Ch}} + p_3^{EB\_Ch}(t)z^{g_3^{EB\_Ch}} + p_4^{EB\_Ch}(t)z^{g_4^{EB\_Ch}}$$
$$= p_1^{EB\_Ch}(t)z^{36} + p_2^{EB\_Ch}(t)z^{24} + p_3^{EB\_Ch}(t)z^{12} + p_4^{EB\_Ch}(t)z^{0}. \quad (8.32)$$

According to the system description one obtains the following output performance stochastic process for the pump controller:

$$\mathbf{g}^{EB\_Pump} = \left\{g_1^{EB\_Pump}, g_2^{EB\_Pump}, g_3^{EB\_Pump}\right\} = \{48,\ 24,\ 0\}$$
$$\mathbf{p}^{EB\_Pump}(t) = \left\{p_1^{EB\_Pump}(t),\ p_2^{EB\_Pump}(t),\ p_3^{EB\_Pump}(t)\right\} \quad (8.33)$$

By solving the following system of differential equations [9]:

$$\begin{cases} \frac{dp_1^{EB\_Pump}(t)}{dt} = -2\lambda^{EB\_Pump} p_1^{EB\_Pump}(t) + \mu^{EB\_Pump} p_2^{EB\_Pump}(t) \\ \frac{dp_2^{EB\_Pump}(t)}{dt} = 2\lambda^{EB\_Pump} p_1^{EB\_Pump}(t) - \left(\lambda^{EB\_Pump} + \mu^{EB\_Pump}\right) p_2^{EB\_Pump}(t) + \mu^{EB\_Pump} p_3^{EB\_Pump}(t) \\ \frac{dp_3^{EB\_Pump}(t)}{dt} = \lambda^{EB\_Pump} p_2^{EB\_Pump}(t) - \mu^{EB\_Pump} p_3^{EB\_Pump}(t). \end{cases} \quad (8.34)$$

under initial conditions

$$p_1^{EB\_Pump}(0) = 1;\ p_2^{EB\_Pump}(0) = p_3^{EB\_Pump}(0) = 0 \quad (8.35)$$

by using MATLAB one obtains a numerical solution for states' probabilities.

Having the sets $\mathbf{g}^{EB\_Pump}$, $\mathbf{p}^{EB\_Pump}(t)$, one defines the L_Z-transform that is associated with the output performance stochastic process of the power controller of the pump:

$$L_Z\left\{G^{EB\_Pump}(t)\right\} = p_1^{EB\_Pump}(t)z^{g_1^{EB\_Pump}} + p_2^{EB\_Pump}(t)z^{g_2^{EB\_Pump}} + p_3^{EB\_Pump}(t)z^{g_3^{EB\_Pump}}$$
$$= p_1^{EB\_Pump}(t)z^{48} + p_2^{EB\_Pump}(t)z^{24} + p_3^{EB\_Pump}(t)z^{0}. \quad (8.36)$$

Using the composition operator Ω_{fser} for the electrical board sub-system consisting of 3 elements that are connected in series, we obtain the L_Z-transform $L_Z\left\{G^{EB}(t)\right\}$ for the electrical board sub-system, where the powers of z in corresponding multiplication are found as a minimum of powers of corresponding terms:

$$
\begin{aligned}
L_Z\{G^{EB}(t)\} &= \Omega_{f_{ser}}\{L_Z\{G^{EB\_Power}(t)\}, L_Z\{G^{EB\_Ch}(t)\}, L_Z\{G^{EB\_Pump}(t)\}\} \\
&= \Omega_{f_{ser}}\Big\{p_1^{EB\_Power}(t)z^{24} + p_2^{EB\_Power}(t)z^0, \\
&\quad p_1^{EB\_Ch}(t)z^{36} + p_2^{EB\_Ch}(t)z^{24} + p_3^{EB\_Ch}(t)z^{12} + p_4^{EB\_Ch}(t)z^0, \\
&\quad p_1^{EB\_Pump}(t)z^{48} + p_2^{EB\_Pump}(t)z^{24} + p_3^{EB\_Pump}(t)z^0\Big\} = \\
&= P_1^{EB}(t)z^{24} + P_2^{EB}(t)z^{12} + P_3^{EB}(t)z^0,
\end{aligned}
\tag{8.37}
$$

where

$$
\begin{aligned}
P_1^{EB}(t) &= p_1^{EB\_Power}(t) \cdot \left(1 - p_3^{EB\_Ch}(t)\right) \cdot \left(1 - p_3^{EB\_Pump}(t)\right), \\
P_2^{EB}(t) &= p_1^{EB\_Power}(t) \cdot p_3^{EB\_Ch}(t) \cdot \left(1 - p_3^{EB\_Pump}(t)\right), \\
P_3^{EB}(t) &= p_1^{EB\_Power}(t) \cdot \left(1 - p_4^{EB\_Ch}(t)\right) p_3^{EB\_Pump}(t) \\
&\quad + p_1^{EB\_Power}(t) \cdot p_4^{EB\_Ch}(t) + p_2^{EB\_Power}(t).
\end{aligned}
$$

8.2.3 *L_Z-Transform for the Entire MRI Water Cooling System Availability*

According to Fig. 8.1, all four sub-systems are connected in series. Therefore, the L_Z-transform that is associated with the entire system output stochastic process $G_s(t)$, can be obtained by applying Ushakov's UGO [7, 10] $\Omega_{f_{ser}}$ over the L_Z-transforms of all sub-systems:

$$
\begin{aligned}
L_Z\{G_s(t)\} &= \Omega_{f_{ser}}\left(L_Z\{G^{Ch}(t)\}, L_Z\{G^{HE}(t)\}, L_Z\{G^{Pump}(t)\}, L_Z\{G^{EB}(t)\}\right) \\
&= \Omega_{f_{ser}}\big(P_1^{Ch}(t)z^{24} + P_2^{Ch}(t)z^{12} + P_3^{Ch}(t)z^0, \\
&\quad P_1^{HE}(t)z^{24} + P_2^{HE}(t)z^{20} + P_3^{HE}(t)z^{16} + P_4^{HE}(t)z^0, \\
&\quad P_1^{Pump}(t)z^{24} + P_2^{Pump}(t)z^0, \\
&\quad P_1^{EB}(t)z^{24} + P_2^{EB}(t)z^{12} + P_3^{EB}(t)z^0\big).
\end{aligned}
\tag{8.38}
$$

After a simple algebra, where powers of z are found as minimum values of powers in corresponding multiplication, the L_Z-transform associated with the output performance stochastic process of the entire system can be obtained:

$$
L_Z\{G_s(t)\} = P_{s1}(t)z^{24} + P_{s2}(t)z^{20} + P_{s3}(t)z^{16} + P_{s4}(t)z^{12} + P_{s5}(t)z^0, \tag{8.39}
$$

where

$$P_{s1}(t) = P_1^{Ch}(t)P_1^{HE}(t)P_1^{Pump}(t)P_1^{EB}(t);$$
$$P_{s2}(t) = P_1^{Ch}(t)P_2^{HE}(t)P_1^{Pump}(t)P_1^{EB}(t);$$
$$P_{s3}(t) = P_1^{Ch}(t)P_3^{HE}(t)P_1^{Pump}(t)P_1^{EB}(t);$$
$$P_{s4}(t) = P_1^{Pump}(t)\left(1 - P_4^{HE}(t)\right)\left(P_1^{Ch}(t)P_2^{EB}(t) + P_2^{Ch}(t)P_1^{EB}(t) + P_2^{Ch}(t)P_2^{EB}(t)\right);$$
$$P_{s5}(t) = \left(P_1^{Ch}(t) + P_2^{Ch}(t)\right)\begin{pmatrix} \left(P_1^{Pump}(t)P_3^{EB}(t) + P_2^{Pump}(t)\right) \\ \left(1 - P_4^{HE}(t)\right) + P_4^{HE}(t) \end{pmatrix} + P_3^{Ch}(t).$$

Using the L_Z-transform (8.36) for the entire system, one obtains MSS instantaneous availability for the MRI cooling system, working under different demand levels: $w_1 = 24$, $w_2 = 20$, $w_3 = 16$ as follows:

$$A_{w=24}(t) = P_{s1}(t);$$
$$A_{w=20}(t) = \sum_{i=1}^{2} P_{si}(t);$$
$$A_{w=16}(t) = \sum_{i=1}^{3} P_{si}(t). \tag{8.40}$$

8.2.4 B-Availability Importance Calculation

Now by differentiating availabilities in expressions (8.37) in accordance with B-availability importance definition (8.3), one will have the B-availability importance expressions [4] for chillers and heat exchangers as functions of time t for different demand levels.

Therefore, accordingly with (8.3) and (8.40) we should find the expressions for B-availability importance of the first chiller as follows:

$$I_{A_{w=24}}^{Ch,1}(t) = \frac{\partial A_{w=24}(t)}{\partial p_{11}^{Ch}(t)} = \left(p_{21}^{Ch}(t) + p_{31}^{Ch}(t) - 2p_{21}^{Ch}(t)p_{31}^{Ch}(t)\right)P_1^{HE}(t)P_1^{Pump}(t)P_1^{EB}(t);$$
$$I_{A_{w=20}}^{Ch,1}(t) = \frac{\partial A_{w=20}(t)}{\partial p_{11}^{Ch}(t)} = \left(p_{21}^{Ch}(t) + p_{31}^{Ch}(t) - 2p_{21}^{Ch}(t)p_{31}^{Ch}(t)\right)\left(P_1^{HE}(t) + P_2^{HE}(t)\right)P_1^{Pump}(t)P_1^{EB}(t);$$

$$
\begin{aligned}
I_{A_{w=16}}^{Ch,1}(t) &= \frac{\partial A_{w=16}(t)}{\partial p_{11}^{Ch}(t)} \\
&= \left(p_{21}^{Ch}(t) + p_{31}^{Ch}(t) - 2p_{21}^{Ch}(t)p_{31}^{Ch}(t)\right) \\
&\quad \left(P_1^{HE}(t) + P_2^{HE}(t) + P_3^{HE}(t)\right)P_1^{Pump}(t)P_1^{EB}(t).
\end{aligned} \tag{8.41}
$$

and B-availability importance of the heat exchangers as following:

$$
\begin{aligned}
I_{A_{w=24}}^{HE,1}(t) &= \frac{\partial A_{w=24}(t)}{\partial p_{11}^{HE}(t)} = P_1^{Ch}(t)p_{21}^{HE}(t)p_{31}^{HE}(t)P_1^{Pump}(t)P_1^{EB}(t); \\
I_{A_{w=24}}^{HE,2}(t) &= \frac{\partial A_{w=24}(t)}{\partial p_{21}^{HE}(t)} = P_1^{Ch}(t)p_{11}^{HE}(t)p_{31}^{HE}(t)P_1^{Pump}(t)P_1^{EB}(t); \\
I_{A_{w=24}}^{HE,3}(t) &= \frac{\partial A_{w=24}(t)}{\partial p_{31}^{HE}(t)} = P_1^{Ch}(t)p_{11}^{HE}(t)p_{21}^{HE}(t)P_1^{Pump}(t)P_1^{EB}(t).
\end{aligned} \tag{8.42}
$$

$$
\begin{aligned}
I_{A_{w=20}}^{HE,1}(t) &= \frac{\partial A_{w=20}(t)}{\partial p_{11}^{HE}(t)} = P_1^{Ch}(t)\left(p_{21}^{HE}(t) + p_{31}^{HE}(t) - p_{21}^{HE}(t)p_{31}^{HE}(t)\right) \\
&\qquad P_1^{Pump}(t)P_1^{EB}(t); \\
I_{A_{w=20}}^{HE,2}(t) &= \frac{\partial A_{w=20}(t)}{\partial p_{21}^{HE}(t)} = P_1^{Ch}(t)p_{11}^{HE}(t)p_{32}^{HE}(t)P_1^{Pump}(t)P_1^{EB}(t); \\
I_{A_{w=20}}^{HE,3}(t) &= \frac{\partial A_{w=20}(t)}{\partial p_{31}^{HE}(t)} = P_1^{Ch}(t)p_{11}^{HE}(t)p_{22}^{HE}(t)P_1^{Pump}(t)P_1^{EB}(t).
\end{aligned} \tag{8.43}
$$

$$
\begin{aligned}
I_{A_{w=16}}^{HE,1}(t) &= \frac{\partial A_{w=16}(t)}{\partial p_{11}^{HE}(t)} = P_1^{Ch}(t)P_1^{Pump}(t)P_1^{EB}(t); \\
I_{A_{w=16}}^{HE,2}(t) &= \frac{\partial A_{w=16}(t)}{\partial p_{21}^{HE}(t)} = 0; \\
I_{A_{w=16}}^{HE,3}(t) &= \frac{\partial A_{w=16}(t)}{\partial p_{31}^{HE}(t)} = 0.
\end{aligned} \tag{8.44}
$$

Calculated B-availability indices are presented in Figs. 8.2, 8.3, 8.4 and 8.5 as functions of time t.

On the Fig. 8.2 one can see that the importance of the first chiller is decreasing over the time and it is maximal for minimal demand level $w = 16$.

The same conclusion can be done from the Fig. 8.3 for the importance of the first heat exchanger—the importance is decreasing over the time and it is maximal for minimal demand.

It is clear that for the demand $w = 24$ all 3 heat exchangers have the same importance, because when any of heat exchangers will fail, the demand $w = 24$ will not satisfied.

For demand level $w = 20$ the importance of the first heat exchanger decreases and importance of the second and the third heat exchangers slightly increases. For

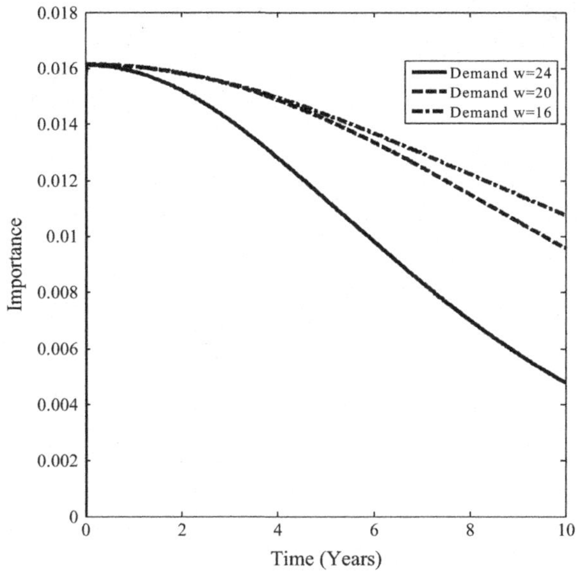

Fig. 8.2 Importance measures $I_{A_{w=24}}^{Ch,1}(t)$, $I_{A_{w=20}}^{Ch,1}(t)$, $I_{A_{w=16}}^{Ch,1}(t)$

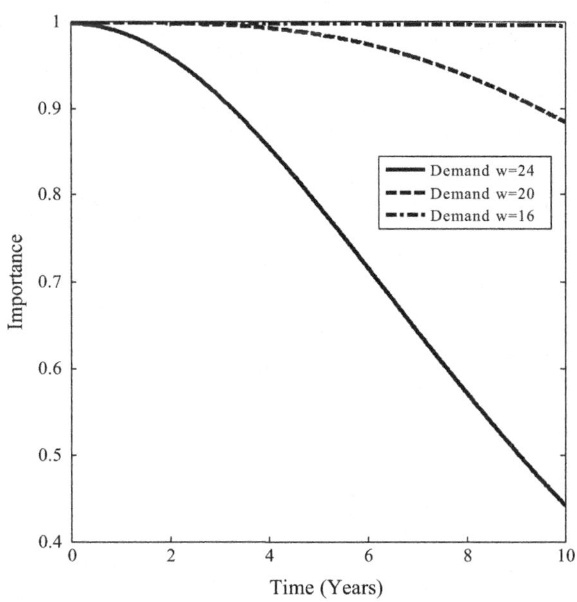

Fig. 8.3 Importance measures $I_{A_{w=24}}^{HE,1}(t)$, $I_{A_{w=20}}^{HE,1}(t)$, $I_{A_{w=16}}^{HE,1}(t)$

demand level $w = 16$ the importance of the first chiller is very high and second and third heat exchangers is equal to zero.

These figures (Figs. 8.2, 8.3, 8.4 and 8.5) evident that in MSS importance for binary elements strongly depends on demand level.

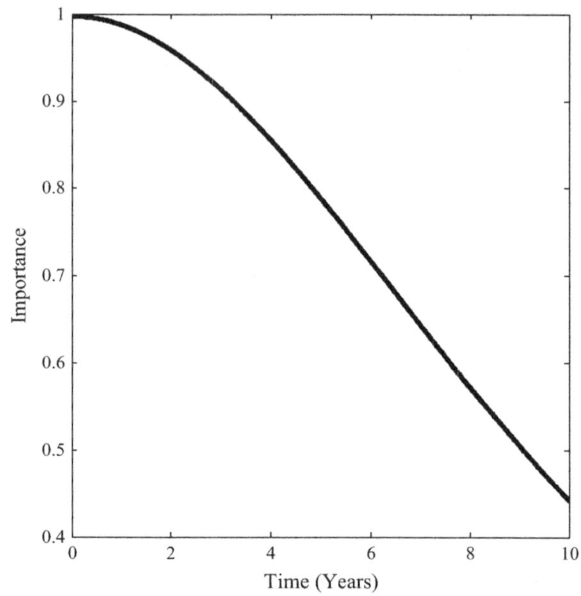

Fig. 8.4 Importance measures $I_{A_{w=24}}^{HE,1}(t)$, $I_{A_{w=24}}^{HE,2}(t)$, $I_{A_{w=24}}^{HE,3}(t)$

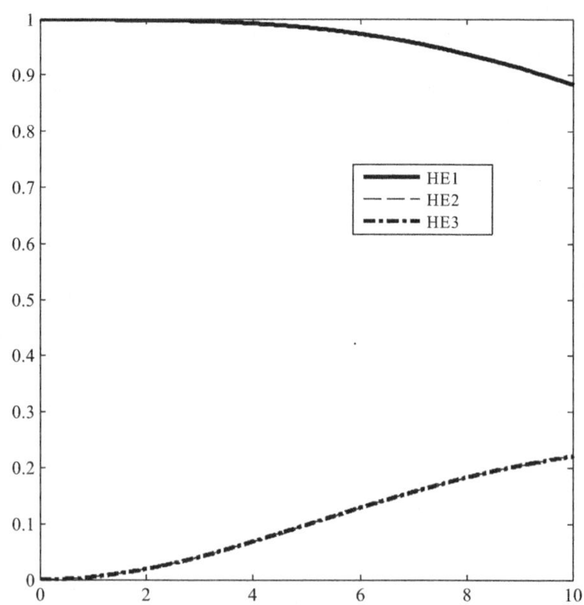

Fig. 8.5 Importance measures $I_{A_{w=20}}^{HE,1}(t)$, $I_{A_{w=20}}^{HE,2}(t) = I_{A_{w=20}}^{HE,3}(t)$

8.3 Summary

In this chapter a new method for computation of the B-availability importance measure for binary state components in an aging MSS under minimal repair was considered. The method is based on the L_Z-transform of the discrete-state continuous-time Markov process.

The proposed method provides a drastic reduction of computational burden compared with a straightforward Markov method or simulation. In the case study considered in the paper, by using a straightforward Markov method, one should be able to build and solve a model with 3840 states. In order to solve this problem by using the proposed method, one should build and solve only 6 simple Markov models with 2 and 3 and 4 states. All other computations are performed by using simple algebra.

Dynamic B-availability importance assessment for aging MSS is especially important, because the relative importance of MSS's components is changing over the time and strongly depends on the system demand.

References

1. Birnbaum Z (1969) On the importance of different components in a multicomponent system. In: Krishnaiah P (ed) Multivariate analysis. Academic Press, New York, pp 561–592
2. Finkelstein M (2008) Failure rate modeling for reliability and risk. Springer, London
3. Frenkel I, Lisnianski A, Khvatskin L (2014) On the L_z-transform application for availability assessment of an aging multi-state water cooling system for medical equipment. In: Frenkel I, Karagrigoriou A, Lisnianski A, Kleyner A (eds) Applied reliability engineering and risk analysis. Probabilistic models and statistical inference. Wiley, Chichester, pp 59–77
4. Kuo W, Zhu X (2012) Importance measures in reliability, risk, and optimization. Principles and applications. Atrium, Wiley, UK
5. Levitin G, Lisnianski A (1999) Importance and sensitivity analysis of multi-state systems using the universal generating function method. Reliab Eng Syst Saf 65(3):271–282
6. Lisnianski A (2012) *Lz*-transform for a discrete-state continuous-time Markov process and its application to multi-state system reliability. In: Lisnianski A, Frenkel I (eds) Recent advances in system reliability. Signatures, multi-state systems and statistical inference. Springer, London, pp 79–95
7. Lisnianski A, Frenkel I, Ding Y (2010) Multi-state system reliability analysis and optimization for engineers and industrial managers. Springer, London
8. Lisnianski A, Frenkel I, Khvatskin L (2015) On Birnbaum importance assessment for aging multi-state system under minimal repair by using the Lz-transform method. Reliab Eng Syst Saf 142:258–266
9. Trivedi K (2002) Probability and statistics with reliability, queuing and computer science applications. Wiley, New York
10. Ushakov I (1986) A universal generating function. Sov J Comput Syst Sci 24:37–48

Appendix
MATLAB Codes for Case Studies Calculations

A.1 Using MATLAB ODE Solvers

For solution of systems of differential equations MATLAB® provide several functions, called *solvers*, that implement Runge-Kutta methods with variable step size. We use in our calculations the `ode45` function, uses a combination of fourth- and fifth-order methods, which is fast and accurate.

When used to solve the vector differential equation $\dot{\mathbf{p}} = \mathbf{f}(t, \mathbf{p})$, specified in the function file `pdot`, whose inputs must be t and p and whose output must be a *column* vector, representing $d\mathbf{p}/dt$; that is, $\mathbf{f}(t, \mathbf{p})$. The number of rows in this column vector must equal the order of the equations. The vector `tspan` contains the starting and ending values of the independent variable t. The vector `p0` contains $\mathbf{p}(t_0)$. The basic syntax is:

```
[t,p]=ode45(@funcpdot,tspan,p0)
```

A.2 MATLAB Code for the Chap. 5—The L_Z-Transform Application for Availability Assessment of Air Conditioning System for Chemical Laboratory

```
%Single Air-Handing Unit blower
function f=funcAHUB(t,p)
global L_AHUB Mu_AHUB;
f=zeros(2,1);
f(1) = -L_AHUB*p(1)+Mu_AHUB*(p(2));
f(2) = L_AHUB*p(1)-Mu_AHUB*p(2);

%Controller
function f=funcC(t,p)
```


A. Lisnianski et al., *Modern Dynamic Reliability Analysis for Multi-state Systems*, Springer Series in Reliability Engineering,
https://doi.org/10.1007/978-3-030-52488-3

```
global L_C Mu_C;
f=zeros(2,1);
f(1) = -L_C*p(1)+Mu_C*(p(2));
f(2) = L_C*p(1)-Mu_C*p(2);

%Single Exhaust blower
function f=funcEB(t,p)
global L_EB Mu_EB;
f=zeros(2,1);
f(1) = -L_EB*p(1)+Mu_EB*(p(2));
f(2) = L_EB*p(1)-Mu_EB*p(2);

%Single Exhaust Hood
function f=funcH(t,p)
global L_H Mu_H;
f=zeros(2,1);
L_Ht=L_H*(1+0.2*t);
f(1) = -L_Ht*p(1)+Mu_H*(p(2));
f(2) = L_Ht*p(1)-Mu_H*p(2);

%Proportional Valve
function f=funcPV(t,p)
global L_PV Mu_PV;
f=zeros(2,1);
L_PVt=L_PV*(1+0.2*t);
f(1) = -L_PVt*p(1)+Mu_PV*(p(2));
f(2) = L_PVt*p(1)-Mu_PV*p(2);

%Solver for MSS Chemical Laboratory

%Variables definition
global L_H Mu_H;
global L_EB Mu_EB;
global L_AHUB Mu_AHUB;
global L_PV Mu_PV;
global L_C Mu_C;

%Variables initialization
L_H = 1.0;
L_EB = 6.0;
L_AHUB = 4.0;
L_PV = 2.0;
L_C = 2.0;
Mu_H = 182.0;
Mu_EB = 90.0;
Mu_AHUB = 90.0;
Mu_PV = 121.0;
Mu_C = 182.0;

%Time initialization
ttt=[0:0.01:1];

%Solution of systems of differential equations
p0_H = [1 0];
[t_H, p_H] = ode45(@funcH, ttt, p0_H);
p0_EB = [1 0];
```

```
[t_EB, p_EB] = ode45(@funcEB, ttt, p0_EB);
p0_AHUB = [1 0];
[t_AHUB, p_AHUB] = ode45(@funcAHUB, ttt, p0_AHUB);
p0_PV = [1 0];
[t_PV, p_PV] = ode45(@funcPV, ttt, p0_PV);
p0_C = [1 0];
[t_C, p_C] = ode45(@funcC, ttt, p0_C);

H_1 = p_H(:,1);
H_2 = p_H(:,2);
EB_1 = p_EB(:,1);
EB_2 = p_EB(:,2);
AHUB_1 = p_AHUB(:,1);
AHUB_2 = p_AHUB(:,2);
PV_1 = p_PV(:,1);
PV_2 = p_PV(:,2);
C_1 = p_C(:,1);
C_2 = p_C(:,2);

% LZ-transforms for SS_HOOD sub-system

SS_HOOD_1 = H_1.^5+5*H_1.^4.*H_2+10*H_1.^3.*H_2.^2;
SS_HOOD_2 = 10*H_1.^2.*H_2.^3;
SS_HOOD_3 = 5*H_1.*H_2.^4;
SS_HOOD_4 = H_2.^5;

% LZ-transforms for SS_EB sub-system
SS_EB_1=EB_1.^2+2*EB_1.*EB_2;
SS_EB_2=EB_2.^2;

% LZ-transforms for SS_AHUB sub-system
SS_AHUB_1 = AHUB_1.^2+2*AHUB_1.*AHUB_2;
SS_AHUB_2 = AHUB_2.^2;

% LZ-transforms for SS1 sub-system
SS1_1 = SS_HOOD_1.*SS_EB_1;
SS1_2 = SS_HOOD_2.*SS_EB_1;
SS1_3 = SS_HOOD_3.*SS_EB_1;
SS1_4 = SS_HOOD_4.*SS_EB_1+SS_EB_2;

% LZ-transforms for SS2 sub-system
SS2_1 = SS_AHUB_1.*SS1_1;
SS2_2 = SS_AHUB_1.*SS1_2;
SS2_3 = SS_AHUB_1.*SS1_3;
SS2_4 = SS_AHUB_1.*SS1_4+SS_AHUB_2;

% LZ-transforms for SS3 sub-system
SS3_1 = PV_1.*SS2_1;
SS3_2 = PV_1.*SS2_2;
SS3_3 = PV_1.*SS2_3;
SS3_4 = PV_1.*SS2_4+PV_2;

% LZ-transforms for CLAB system
CLAB_1 = C_1.*SS3_1;
CLAB_2 = C_1.*SS3_2;
CLAB_3 = C_1.*SS3_3;
```

```
CLAB_4 = C_1.*SS3_4+C_2;

% Plot of Instantaneous availability of Laboratory Hoods
Subsystem
%plot(ttt, H_1,'k-',ttt, SS_HOOD_1,'k-','LineWidth',2);

%Plot of Instantaneous availability of Exhaust Blower Subsystem
%plot(ttt, EB_1,'k-',ttt, SS_EB_1,'k-','LineWidth',2);

%Plot of Instantaneous availability of Air Handing Unit Blowers
Subsystem
%plot(ttt, AHUB_1,'k-',ttt, SS_AHUB_1,'k-','LineWidth',2);

%Plot of Instantaneous availability of Proportional Valve
%plot(ttt, PV_1,'k-','LineWidth',2);

%Plot of Instantaneous availability of Controller
%plot(ttt, C_1,'k-','LineWidth',2);

%Plot of Availability of Air Conditioning System
plot(ttt, CLAB_1,'k-','LineWidth',2);

ylabel('Availability','FontSize',12,'FontWeight','bold',
'FontName','Times New Roman');
xlabel('Time (Year)','FontSize',12,'FontWeight','bold',
'FontName','Times New Roman');
set(gca,'FontSize',12,'FontWeight','bold','FontName',
'Times New Roman');
```

The manufacturer's authorised representative in the EU is Springer Nature Customer Service Centre GmbH, Europaplatz 3, 69115 Heidelberg, Germany. If you have any concerns regarding our products, please contact ProductSafety@springernature.com

Printed and bound by CPI Group (UK) Ltd, Croydon, CR0 4YY

30/09/2026

02225505-0001